THE DESIGN AND COMPLEXITY OF THE

THE DESIGN AND COMPLEXITY OF THE

JEFFREY P. TOMKINS

The Design and Complexity of the CELL
by Dr. Jeffrey P. Tomkins

Dr. Jeffrey Tomkins earned a master's degree in Plant Science in 1990 from the University of Idaho, where he performed research in plant hormones. He received his Ph.D. in Genetics from Clemson University in 1996. While at Clemson, he worked as a research technician in a plant breeding/genetics program, with a research focus in the area of quantitative and physiological genetics in soybean. After receiving his Ph.D., he worked at a genomics institute and became a faculty member in the Department of Genetics and Biochemistry at Clemson. He had become a Christian as an undergraduate at Washington State University in 1982, with a goal to eventually work as a scientist and author in the creation science field. In 2009, Dr. Tomkins joined the Institute for Creation Research as Research Associate.

Contributing authors: Dr. Nathaniel Jeanson, Dr. Henry Morris, Dr. Brad Forlow, Dr. Randy Guliuzza, Mr. Brian Thomas, and Mr. Frank Sherwin

First printing: June 2012

Cover Design: Dennis Davidson
Interior Design: Susan Windsor

All Scripture quotations are from the King James Version.

ISBN: 978-1-935587-08-8
Library of Congress Catalog Number: 2012939464

Please visit our website for other books and resources: www.icr.org.

Printed in the United States of America.

Table of Contents

Page

Contributors **7**

Foreword **9**

Introduction **10**

Chapter 1 Dogma and Science Surrounding Cell Origins **13**
- Abiogenesis 13
- The Evolutionary Problem of the Cell 17
- Have Scientists Created Artificial Life? 18
- Summary 19
- Application: Have Scientists Created a Synthetic Cell? 20

Chapter 2 A Basic Description of Cells and Cell Types **23**
- A Basic Definition of Cells 23
- Two Basic Cell Types in Biology 25
- Prokaryotic Cell Characteristics 27
- Development of a Model System for Cell Biology 30
- *E. coli*: The Prokaryotic Model 30
- The Role of Cells in Biological Creation 31
- Summary 33
- Application: Life's Indispensable Microscopic Machines 34

Chapter 3 Eukaryotic Cells and Multicellularity **37**
- The Highly Engineered Eukaryotic Nucleus and Mitochondria . . 38
- Mitochondria: The Evolutionary Model of Symbiosis Is Unsupported . . 40
- Plant Cell Chloroplasts 41
- Additional Eukaryotic Cell Characteristics 42
- Summary 45
- Application: Life-Giving Blood 46

Chapter 4 Genes and Proteins **49**
- DNA: The Cell's Genetic Blueprint 49
- Making a Protein 50
- A Masterpiece of Design and Purpose 53
- Summary 54
- Application: Where Did Flesh-Eating Bacteria Come From? . . . 55

Chapter 5 Cell Division and DNA Replication: How Life Is Engineered to Perpetuate **57**
- Cell Division: The Basic Biological Process Required for Life . . 57
- Mitosis 59
- Meiosis 60
  - Meiosis I 63
  - Meiosis II 63
- Summary 64
- Application: Brain's Complexity "Is Beyond Anything Imagined" . . . 65

Chapter 6 **Cell Signaling: The Miracle of the Biological Network** **67**
Cell Surface Receptors 67
Neurons 69
Summary 71
Application: Human Reproduction 72

Chapter 7 **The Cytoskeleton and the Extracellular Matrix: How Biology Achieves Shape, Form, and Movement** **75**
Actin Filaments and Microtubules 75
Junctions between Cells 77
The Extracellular Matrix 79
Summary 80
Application: The Mysteries of Stunning Soft Tissue Fossil Finds 81

Chapter 8 **Cells in Creation, Sin, and Redemption** **83**
Plants 83
Photosynthesis and Respiration 84
Water, Air, and Land Creatures 85
Consequences of the Curse 86
Summary 87
Application: Immune Systems, the Body's Security Force 88

Chapter 9 **The Biology of Stem Cells** **91**
The Biology of Adult Stem Cells 91
Cellular Mechanisms of Self-Renewal and Differentiation 93
Molecular Mechanisms of Self-Renewal and Differentiation 94
The Biology of Embryonic Stem Cells 95
How Embryonic Stem Cells Work: Mechanisms of ESC Self-Renewal and Differentiation 95
The Origins of Human Cellular Diversity: The Fact of Creation and the Insufficiency of Evolution 96
Summary 97
Application: Darwinian Medicine: A Prescription for Failure 98

Chapter 10 **Processes and Implications of Stem Cell Research** **101**
Reprogramming and Cloning 103
Stem Cells and Medicine: Cell Therapy 105
Stem Cell Ethics 106
Conclusion 109
Summary 110
Application: Cells: Sophisticated and God-Designed 111

Appendix 1 **Biology and the Bible** **113**

Appendix 2 **The Development of Pharmaceutical Therapeutics** **117**
Therapeutic Intervention 117
Pharmaceutical Drug Discovery and Development 118
Drug Discovery (Preclinical R&D) 118
Drug Development (Clinical R&D) 121
Drug Approval and Commercialization 122
Pharmacological Inhibition 123
Conclusion 124

Glossary **125**

Credits **128**

Index **129**

Contributors

Nathaniel T. Jeanson, Ph.D.
Deputy Director for Life Sciences
Institute for Creation Research

- Chapter 9: The Biology of Stem Cells
- Chapter 10: Processes and Implications of Stem Cell Research

Henry M. Morris, Ph.D.
Founder
Institute for Creation Research

- Appendix 1: Biology and the Bible

Brad Forlow, Ph.D.
Associate Science Editor
Institute for Creation Research

- Appendix 2: The Development of Pharmaceutical Therapeutics

Randy J. Guliuzza, P.E., M.D.
National Representative
Institute for Creation Research

- Life's Indispensable Microscopic Machines
- Life-Giving Blood
- Human Reproduction
- Immune Systems, the Body's Security Force
- Darwinian Medicine: A Prescription for Failure

Brian Thomas, M.S.
Science Writer
Institute for Creation Research

- Have Scientists Created a Synthetic Cell?
- Where Did Flesh-Eating Bacteria Come From?
- Brain's Complexity "Is Beyond Anything Imagined"
- The Mysteries of Stunning Soft Tissue Fossil Finds

Frank Sherwin, M.A.
Research Associate, Senior Lecturer, and Science Writer
Institute for Creation Research

- Cells: Sophisticated and God-Designed

Foreword

The functions within the cells of our bodies are foundational to our existence. Understanding these functions has made the environment and the processes of our lifestyles healthier, more enjoyable, and more productive. All of humanity has benefited from the life sciences and the scientists who have dedicated their considerable skills to uncovering the functions and processes of the multifaceted variety of cells.

Although scientists have discovered, documented, and developed wonderful insights about the complex information, precise sequential processes, and unique interwoven controls within cells, there is a huge chasm among scientists when they try to understand how these highly efficient processes got started in the first place.

The majority of scientifically trained biologists and geneticists are taught that the *apparent* design that is observed in cells is the result of only random chemical and energy processes operating over eons of time. Such a belief system prompts their thinking to rest on purely natural logic, producing materialistic conclusions—and often ignoring or marginalizing the implications of careful engineering and design.

There are, however, multiple thousands of scientists who accept the Bible's message that there is a Creator who planned the creation, designed the intricate engineering efficiencies into that creation, and then built the product: our planet and its wonderfully unique life and functions. That belief system not only fits empirically with what is observed (design, precise function, operational efficiencies, etc.), but provides insight that enables creation-based scientists to grasp the significance of the information more readily—without having to invent a supposed eons-long story for the development of what is actually observed.

Dr. Jeffrey Tomkins and his contributing colleagues have provided an excellent resource that will document and help explain the intricate processes of cells and give keen insight for "clearly seeing" the obvious hand of the Creator in the "things that are made" (Romans 1:20).

Each chapter is written for technical accuracy. Most high school biology students will find the information similar to what is in their textbooks—but more carefully explained regarding the clear presence of design. Interspersed throughout the book are short articles that will provide specific examples of the cellular functions as they impact system operation. These examples will provide observable applications of how these marvelous processes insure that the purpose of each design is fulfilled.

At the end of each chapter is an information summary that will reinforce the science discussed and outline the design characteristics easily identified by the discoveries.

This book is designed as a scientific resource and ready reference, as well as an apologetic tool to use as a witness of the omnipotent and omniscient Creator and Savior. Our prayer at the Institute for Creation Research is that you will find both of these purposes fulfilled in your life.

Henry M. Morris III
Chief Executive Officer
Institute for Creation Research

Introduction

The growing field of biology—including biomedical, agricultural, and environmental sciences—is becoming more prevalent in our world and more relevant to our daily lives.

This field is dominated by a strong philosophical component that plays an important role in our lives. That component is *worldview*. The worldview of evolutionary naturalism impacts the moral fabric of our society, and we encounter its influences on a daily basis. The majority of the scientific establishment is sold on the belief that there is no God or Creator and that life developed spontaneously and randomly. On the other hand, many Christians are unaware that recent scientific discoveries in the area of biology actually support an opposite worldview, one that validates special creation as recorded in the Bible. Despite the predominance of evolutionary thinking in education and science, Christians can have confidence that what God said about creation, and particularly about life, is absolutely true.

Although the scientific establishment has sought to suppress much of the evidence for creation, credentialed scientists within the creation and intelligent design (ID) movements have successfully brought this evidence to light. For instance, the book that started the modern creation science movement in 1961—*The Genesis Flood* by John Whitcomb and Henry Morris—demonstrated the geological evidences that confirm the historical global Flood account of Genesis. A number of recent popular books that incorporate new biological discoveries have made *The New York Times* bestseller list, such as *Signature in the Cell* by science philosopher Stephen Meyer.

The bottom line is that understanding the basics of biology and its support for the biblical account of creation will not only build your own faith, but it will also provide a powerful tool for evangelism and the defense of the faith.

Why another book on biology? If the science community is run by atheists and agnostics, why bother to study? And if there are good books on biology by those in the ID movement, what else can be said about creation and biology?

While many of the ID books make a case for a designer (through inference), these books fail to point readers to the Bible and the Creator. There is no doubt that the ID movement has helped pave the way for some to rethink their belief in evolution. However, ID proponents often retain belief in evolutionary ideas—the Big Bang, millions and billions of years, common ancestry, etc.

Our goal in this book is not only to bring you up to speed on basic biology—in this case, the design and complexity of the cell—but also to address the various evolutionary arguments that have dominated and shaped the academic environment of the early 21st century. Knowing the facts about the cell is important; God is to be praised for creating life with such attention to detail. However, knowing and using these facts to defend your faith or to persuade others to consider the claims of the Creator will provide you with the ammunition you need to counter the various evolutionary arrows that will surely come your way. This is especially true for those in or on their way to college. Sadly, evolutionary naturalism in the biological sciences is just as likely to be accepted and taught at a supposedly Christian college as it is at a state-supported secular school.

Additionally, this book will prepare you to have educated conversations on these topics. For example, what issues are involved in stem cell research, and why do they matter? The whole controversy about cloning makes the news often. Is cloning animals okay? Why not human beings?

As we read the news about new discoveries, how can the average person evaluate the truth behind the headlines? An understanding of these topics is vital, as many touch on our health and that of our children. Furthermore, as Christians, we should be prepared to give an intelligent, Bible-based response to the issues that challenge our belief in the facts of Scripture or that seek to turn our belief away from the Creator.

Science is not a morally neutral discipline. There is a worldview that governs the beliefs and actions of those who do science. If science is not guided by strong moral and biblical values, the selfish and evil sin nature of man will turn our world into a technological nightmare for humanity.

Because of the natural progression of this book, with one topic building upon another, we recommend that you not jump around from chapter to chapter. In addition, keep this book as a resource, not only for yourself as you understand the building blocks of life in greater detail, but also as you hone your skills to defend your faith. By the end of this book, you should better understand the wonder of creation found in the marvelously designed cell, and thereby honor the Creator of all life.

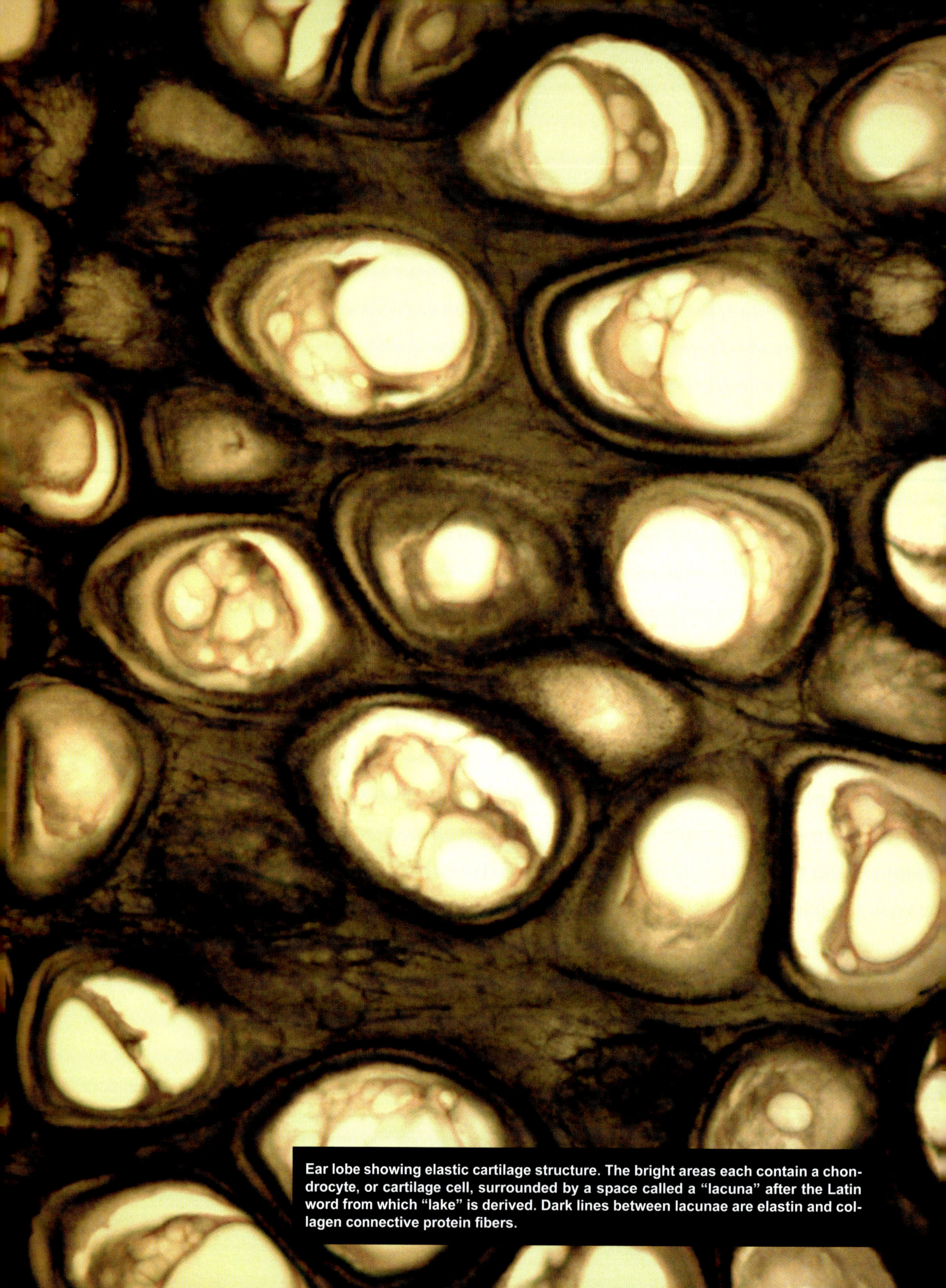

Ear lobe showing elastic cartilage structure. The bright areas each contain a chondrocyte, or cartilage cell, surrounded by a space called a "lacuna" after the Latin word from which "lake" is derived. Dark lines between lacunae are elastin and collagen connective protein fibers.

Chapter 1

Dogma and Science Surrounding Cell Origins

The biggest challenge to evolution is perhaps the origin of the first cell and its associated biomolecules. In fact, many evolutionists have completely abandoned research in this area simply because it has been exceedingly unfruitful for their cause. Even with the most ingenious engineering of equipment and conditions to supposedly simulate some sort of early evolutionary event involved in the formation of the first particles of life, these studies have either completely failed or have produced entirely unsupportive results. And yet, most of the biology science community ignores this complete lack of evidence for a naturalistic origin of life, considers it a part of their dogma, and has gone on to pursue other avenues of research that are more achievable.

Abiogenesis

Abiogenesis is the study of how cellular life supposedly began with the development of the first biomolecules and the first primitive cell from previously inanimate chemicals. One textbook described the mythical process this way:

> The first true "cells" arose when DNA, RNA, and proteins became contained within a boundary, the plasma membrane. These first unicellular organisms acquired the ability to interact. Multicellularity endowed these early organisms with the ability to organize into ever more complicated structures, ultimately giving rise to the explosion in biodiversity we see today.[1]

Scientists point out that abiogenesis is different from the study of evolution, although both are supposedly governed by random naturalistic processes. According to the current dogma, evolution refers to how life progressed and diversified following abiogenesis. The abiogenesis event would have had to involve at least three primary areas: 1) formation of large biomolecules, along with a system for reproducing them (replication); 2) the organization and enclosure of these molecules within a functional membrane-bound system; and 3) the development of the first metabolic processes needed for energy production to feed the system and keep it alive. Except for the third part of the event, scientists have proposed hypothetical models for how these things may have occurred.

In research related to the spontaneous generation of biomolecules, scientists have made the most progress by experimenting with amino acids, which are the building blocks of proteins. Research with RNA has also been conducted. However, no studies have provided any evidence for the spontaneous generation of nucleotides, the building blocks of RNA.

The "RNA world" idea involves a scenario in which the first information-containing molecules needed to create the first cell genome were composed of RNA, not

1 Corbett, S. A. and R. A. Foty. 2008. Cell Structure, Function, and Genetics. In *Surgery: Basic Science and Clinical Evidence*, 2nd ed. J. A. Norton et al, eds. New York: Springer, 37.

DNA. This whole concept came about because of the need to have a molecule that could not only contain informational instructions or code, but could replicate itself and even perform other catalytic functions. In the typical cell scenario, genetic information in the form of a code is carried in the DNA molecule, just like the hard drive on a computer carries coded information for computer programs. When the cell wants to make a protein, a copy of a gene (part of the DNA code) is made using another type of nucleic acid, appropriately called a transcript or messenger RNA. This is similar to a computer program being copied from the hard drive and then placed in some sort of temporary memory like RAM (random access memory). The genetic code is never executed directly from DNA, just like the operational code in the computer is never directly executed from the hard drive. Instead, a copy of the code is made, stored in memory, and then executed.

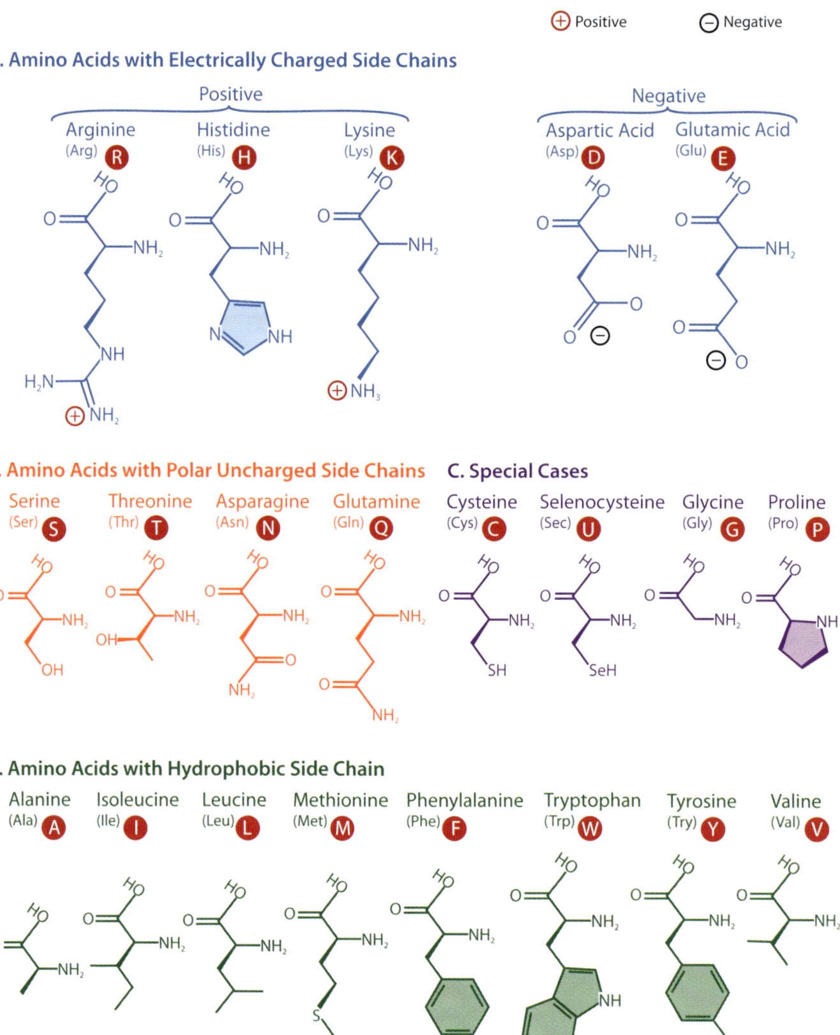

Figure 1.1 Amino acids used in biological systems and their chemical properties.

There are many mysteries and wonders of the cell, and metaphors and analogies involving computers are used in this book because, just like the cell, a computer is an engineered system that features many of the same principles of design. Of course, cells and biological systems are infinitely more complex than computers, but many of the elementary design principles regarding the flow of information and its usage are similar. It is no wonder that the computer science industry currently has such a strong fascination with how DNA and the genome work. The most successful inventions of modern times are usually based on mimicking newly discovered mechanisms found in God's creation.

In a cell, the RNA transcript is used in conjunction with another type of cellular machinery, called a *ribosome*, to construct a specific and exact sequence of amino acids needed to create a certain type of protein. There are about 20 different amino acids with a wide variety of chemical properties that are used to construct different proteins (Figure 1.1). Some proteins contain literally hundreds of amino acids that must be ordered perfectly or else the protein will not function properly. The typical information flow in the cell is DNA to RNA to protein and has been termed the "Central Dogma" (Figure 1.2). DNA and RNA are needed to make protein, protein and RNA are needed to make more DNA, and DNA and protein are needed to make RNA. This three-way interdependency is even more confounding to the evolutionist and the case for abiogenesis than the traditional "chicken or the egg" scenario.

The reason that the "RNA world" idea gained strength as a model for abiogenesis is because some RNA molecules are used directly for a functional purpose in the cell and not as an intermediate transcript to produce a protein. In fact, some RNAs have basic catalytic activity, breaking and forming simple bonds in nucleic acids. However, this is done generally in conjunction with one or more proteins and occurs as part of an intricate process involving a variety of other biomolecules. At

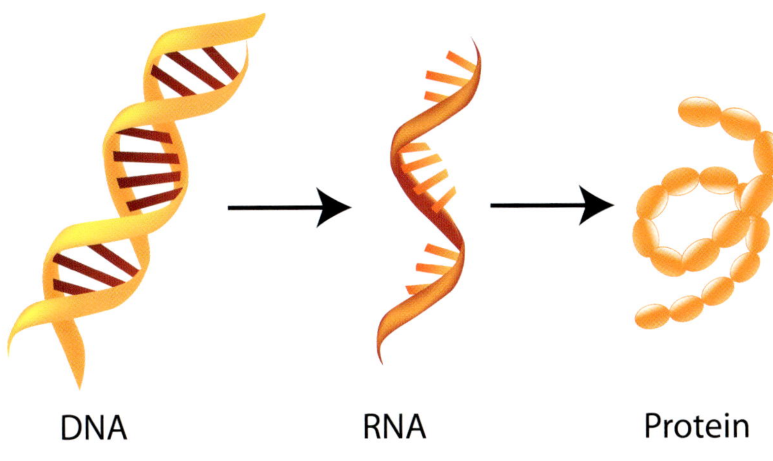

Figure 1.2 The so-called "Central Dogma" of molecular biology. This represents an oversimplified perspective on the general flow of information in the cell. In the cycle of interdependency, both DNA and RNA require specialized proteins for their production and utilization. Proteins themselves require other specialized proteins for their production, in addition to DNA and RNA.

present, RNA has never been observed to replicate itself, or any other molecule, for that matter.

Another feature that makes RNA unique—and also helps to eliminate it as the first spontaneously produced biomolecule—is that it typically occurs as a single-stranded nucleic acid. The RNAs that foster some sort of catalytic activity do so by looping back on themselves and base-pairing with complementary sequences (see chapter 4). Thus, they assume unique shapes called secondary structures. The RNAs that perform this sort of task without being bound in some protein are very short, and for good reason. Because RNAs are single-stranded, they are very unstable, especially when they are not present in their highly engineered and protected cellular environment. The DNA molecule occurs in a double-stranded form and is packaged with a variety of protective proteins, making it considerably more stable. Thus, DNA has been chosen as the material for use in lipid packaging experiments that supposedly show how a first cell may have formed.

The first stage in abiogenesis must be the formation of key biomolecules needed for cell formation. Early abiogenesis research focused on the formation of biomolecules under conditions that emulated the supposed primitive environment of early earth. The prevailing view of this model was first postulated by Aleksandr Oparin and came to be called "the warm soup theory."

In general, it is believed that the first origins of life occurred in the oceans within a biochemical soup that supposedly once existed. The precursor molecules were thought to have been simple organic compounds that were catalyzed to form larger biomolecules through energy sources such as lightning, ultraviolet light, deep-sea hydrothermal vents, volcanoes, meteorites, hot springs, and electrical discharges from the sun. In fact, this general model, which has several variants, is still the most widely held view on the subject. However, it is adhered to not because it is scientifically valid—virtually all experiments performed in this area have produced negative results—but rather because it is currently the only alternative to special creation.

One of the first and perhaps most famous tests of the warm soup theory was the Miller-Urey experiments. While some evolutionists have labeled these experiments a success, many creationists have shown that the experiments actually served to reinforce the biblical model of creation, demonstrating that biomolecules such as proteins could not arise by naturalistic random processes, but instead require a creator—a divine engineer.

The Miller-Urey experiment involved feeding a large glass chamber with a gaseous mixture that contained organic compounds (methane, ammonia, water vapor, and hydrogen) thought to be necessary for the generation of biomolecules (see Figure 1.3). A liquid mixture containing the organic compounds was kept in a constant gaseous state by heating and boiling condensed liquids at the bottom of an adjacent flask that had a tube to feed the gas into a reaction chamber. A tungsten sparking device capable of emitting a discharge of about 60,000 volts (simulating a lightning strike) was placed in the gas chamber to stimulate the formation of new compounds. Below the sparking chamber was a condensation trap where the reactants could be collected and sampled. The experiment was allowed to progress for several days. When it was terminated, a tar-like residue was collected, but the attempts never contained real proteins. After some modification of the reaction conditions, trace amounts of a few basic amino acids were obtained.

Amino acids are the base molecules that form proteins. In most living things, proteins are formed from about

20 different amino acids. Proteins are essentially chains of amino acids that are formed into various three-dimensional shapes and are often bound to other molecules (e.g., sugars, metal ions, lipids, or other proteins). In the cell, proteins are the principle biomolecules that perform most of the work, and they also form the key structural features in the cell's framework.

The amino acids that Miller-Urey obtained in their experiment were mostly glycine and alanine, the most simple of all the amino acids. Following this experiment, additional experiments performed with improved engineering of the devices and new conditions resulted in the production of small amounts of several other simple amino acids. When this phase of research had progressed to the point where no further progress could be made, scientists had only been able to produce fewer than half of the amino acids needed to make proteins. The production of more advanced types of amino acids required extremely complex conditions and apparatus, far beyond what scientists thought were early earth conditions. Of course, many would argue that the highly engineered apparatus Miller-Urey and others used was a far cry from random naturalistic processes and conditions.

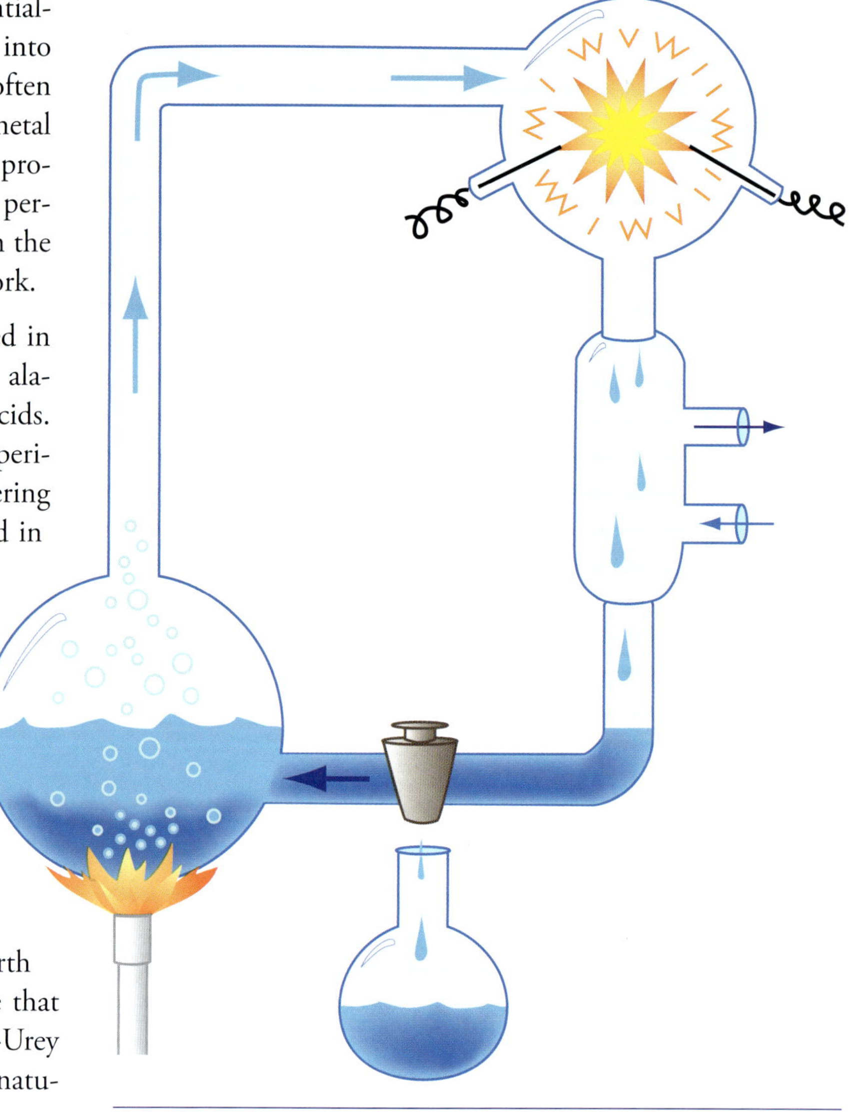

Figure 1.3 Apparatus used in Miller-Urey amino acid genesis experiments.

There are a few key issues that are noteworthy in regard to these experiments. First, the experimental conditions required the absence of oxygen. Earth's current atmosphere contains oxygen that otherwise would have destroyed the created compounds through a chemical process called oxidation. The carefully engineered oxygen-free experiment contained a "reducing environment," which is claimed to have been the same type of atmosphere that existed on the early earth. The problem with this hypothesis is that we presently live in an oxygen-rich environment, and geologists claim, based on the condition of various types of rocks, that the earth has always been an oxygen-rich or oxidizing environment.

Second, the amino acids that were produced in the reaction chambers were a *racemic* mixture, occurring in both right and left-handed forms. The best analogy for this chemical scenario is a person's hands. A human has two hands that are essentially identical except for the fact that one is a mirror image of the other, producing a left and right-oriented version of each one. Biological systems are designed to only produce and utilize left-handed amino acids to make proteins. The presence of a racemic mixture would have been prohibitive to the formation of functional proteins.

Third, the experimental conditions that Miller-Urey used denature and destroy proteins, preventing the formation of functional proteins. While these conditions may have produced a few simple amino acids, they were certainly not conducive to their polymerization for protein formation. In fact, even if all the amino acids needed for life were present in only a left-handed form and existed in adequate concentrations, and even if the conditions did not destroy proteins, there would have existed no mechanism to polymerize or connect them and make proteins. The production of proteins is a highly complex and ordered polymerization process that requires numerous other proteins and various other molecules (e.g., transfer RNAs).

In addition, there must also exist an information source that tells the cellular machinery which amino acids to use and in what order they are to be placed. Amino acids are divided into groups based on their chemistries; some are hydrophobic (repel water) and others are hydrophilic (attract water). Amino acids also have different charges—positive, negative, and neutral. Finally, although all amino acids have the same basic backbone structure, the properties of amino acids are derived from their unique side-chains. Without the correct information directing the assembly of a protein, it would be impossible for a functional version that served some useful purpose in the cell to be generated. The problem of functional proteins and a required information source is as much of a roadblock to abiogenesis as the problem of producing some twenty different amino acids in the right concentration and in the right orientation.

Even though the Miller-Urey experiments produced insufficient results, these results inspired Sydney Fox to produce a series of experiments with the goal of polymerizing amino acids into polypeptide chains or proteins. The original experiments by Fox used extremely high heat to cause the amino acids to coalesce and form globular ring-like clusters, which Fox called *proteinoids*. This result is a far cry from anything resembling a real protein, which is a chain of amino acids connected by peptide bonds.

Later, the technique used to form proteinoids was refined by replacing the heat variable of the experiment with a strong acid. Many scientists felt that the use of an acid was a closer resemblance to primitive earth conditions at the time that biomolecules supposedly formed. Once again, as with the amino acid formation experiments by Miller-Urey, there were a number of very serious problems with these experiments that rendered them generally unproductive in explaining abiogenesis.

First, Fox started with a mixture of left-handed amino acids derived from real biological systems, not the racemic mixture of simple amino acids created in the Miller-Urey experiments. In addition, the molecules Fox formed were small, ring-shape groups of random amino acids and did not remotely resemble real proteins that are polymerized via peptide bonds in a linear format based on specific genetic information. Like the Miller-Urey experiments, the reaction products had to be removed immediately from the system or they would have been destroyed by the same highly engineered conditions that generated them.

While the Miller-Urey and Fox experiments did produce marginal results—although completely unsupportive of abiogenesis—experiments by scientists to produce nucleic acids and lipids via chance random naturalistic processes fared even worse.

Cells are composed of many types of complex biomolecules that function together in unison. It is not reasonable to have one or two classes of chemicals spontaneously appear, wait around for additional biomolecules to spontaneously form, and then have them all self-assemble into something representing a primitive cell. Interestingly, the nucleic acids needed to store genetic code (protein assembly instructions) and the lipids needed to provide the external and internal membranes for the cell are the primary biomolecules that so far have completely resisted formation in any engineered so-called abiogenesis experiment.

The Evolutionary Problem of the Cell

Not only is the naturalistic origin of the first biomolecules (DNA, RNA, proteins, and lipids) an insurmountable hurdle for evolution to overcome, but the supposed origin of the first cell is an even greater obstacle. Even if evolutionary scientists are given the benefit of the doubt and granted the possibility that somehow complex biomolecules were able to form spontaneously, the formation of these into the first cell is an even more immeasurably complex step.

According to the current general model of abiogenesis, several naturalistic spontaneous occurrences had to happen at or near the same time to create something that led to the first cell. Based on basic chemical principles, this is not possible. But assuming a nucleic acid of some sort was able to manifest itself from the primordial soup, what would be the next step to pave the way for the first cell?

Currently, the most popular model is that RNA or DNA and some form of self-assembling simple lipids arose as the first precursor to a cell. It is thought that the lipids formed a primitive membrane and encased a nucleic acid chain (polymer) that would have formed the first cell genome. Scientists are now experimenting with this scenario using DNA fragments, because RNA is a highly unstable molecule and difficult to work with outside the cell. The problem with this line of research is that it does not fit well with the supposed "RNA world" idea that is quite popular among scientists today.

The first factor that must be considered in such a scenario is the information required to bring about a truly functioning simple cell. Even just a simple bacterium like *Escherichia coli (E. coli)* contains a little over 4,400 genes that specify how it is to be constructed, physiologically maintained, and propagated. The lowest number of genes ever found in a bacterium was 468, which was discovered when the genome of the parasitic bacterium *Mycoplasma genitalium* was sequenced in the mid-1990s. Of course, this bacterium was not self-sufficient and would never survive outside of an animal host. Nevertheless, scientists used data like this from a variety of simple bacteria to postulate and produce what they called a minimal set of 256 genes required for basic cellular life. The hypothetical reasoning is that this minimal gene set would be enough for the survival of a bacterium that represents a basic cell supposedly similar to one of the first cells produced during abiogenesis and evolution.

The fact of the matter is that even the simplest bacterial cell is exceedingly complex. Scientists have, for the most part, abandoned research in abiogenesis, particularly in regard to formation of the first cell and how it might have occurred. Even though the extensive research in prebiotic chemistry yielded very little, if any, support for abiogenesis, the huge theoretical gap between the first simple biomolecules and the formation of the first cell has never been bridged with any coherent hypothetical model. The formation of the first cell is just too complicated an experiment to achieve, even with the best laboratory engineering, a topic we will address in the next section.

Perhaps one of the most productive pieces of research in abiogenesis was performed in the early 1980s by Sir Fred Hoyle, a mathematician and astronomer in England. In a groundbreaking study determining the mathematical probability of life arising by chance, Hoyle estimated the probability for the required set of enzymes for even the simplest living cell to arise by chance random processes. He calculated a statistic of one in $10^{40,000}$ (one followed by 40,000 zeroes!). Since the hypothetical estimate of the number of atoms in the known universe is quite small by comparison (10^{80}), even a whole universe full of primordial soup would not have the slightest chance. Shortly after these findings, Hoyle used a very enlightening comparison to clarify the improbability of even the simplest cell arising by chance. He likened it to the same probability as a hurricane going through a junk yard and leaving in its wake a Boeing 747, full of fuel and ready to fly. Surprisingly, Hoyle never stated a belief in special creation despite all of the overwhelming evidence that he himself generated while being a devout skeptic and atheist. He died in 2001 while still searching for some solution to this evolutionary dilemma.

Have Scientists Created Artificial Life?

The research concerning the basic complement of genes that are required for the most basic of bacterial life paved the way for modern researchers in cell biology to eventually claim that they had created artificial life. But is this really the case?

It is true that the scientists fabricated a synthetic circular bacterial chromosome by splicing together segments of DNA synthesized in the lab. They then inserted this man-made chromosome into an already functional bacterial host cell. This DNA synthesis technology is not new and has been available since the mid-1990s. In fact, there are companies that do this regularly and charge by the nucleotide base. However, the length of sequence that can be produced artificially is limited, so in the case of a large circular bacterial chromosome, synthesized DNA fragments had to be spliced together to form a whole bacterial chromosome. The trick is being able to maintain and manipulate such a large DNA molecule without physically breaking it into smaller fragments. This laboratory technology was perfected by a group of scientists at the J. Craig Venter Institute near Washington, DC.

The larger question is this: Did these scientists really create artificial life?

Where did they get the DNA code for the bacterial genes to make their "so-called" artificial chromosome? They acquired the information from *pre-existing* DNA sequence taken from the genes of bacteria whose DNA was sequenced. And to add a little more human arrogance to the project, they also coded the names of lab workers into the DNA so as to not disrupt the function of the original DNA sequence created by God and discovered thousands of years later by scientists. They called this added useless genetic code *DNA watermarks*. In the end, all that the researchers really accomplished was the development of advanced laboratory techniques that exploited and manipulated pre-existing biological information and material. In fact, scientists do not have the capability to artificially create a fully functional bacterial cell from scratch. This type of capacity is still way beyond the current technology.

Of course, the media exploited these results as proof of evolution and announced that life can occur without God. Experiments like this actually highlight the fact that engineering and creativity are always involved in working with and manipulating even the simplest forms of life.

Summary

1. *Abiogenesis* is the study of how cells could have developed from nonliving matter. Studies on the origin of life depend heavily on the presuppositions used. A belief in evolutionary naturalism will direct the answers one way, and a belief in a Creator will direct the answers in a very different way.
2. Many experiments have been conducted to try to produce simple biomolecules from various hypothetical "early earth" environmental "soups." None have produced anything even close to a functioning cell. The trace amounts of so-called "building blocks of life" that have resulted from these experiments have merely demonstrated that the best scientific minds in the world, using the most sophisticated laboratory equipment and controlled environments, can at best only produce crude copies of what was already here in the first place. And those copies are a far cry from the many types of complex biomolecules required to support cellular life.
3. All cells display evidence of engineering and design. It is legitimate to compare the functions of a cell to a computer, a machine, or a complex building project. Many successful human inventions are based on mimicking newly discovered biological features found in God's creation.
4. The "Central Dogma" of cellular process recognizes that *information* in the cell *flows* from the DNA to the RNA to the protein. The origin of this three-way interdependent process cannot be explained using random evolutionary ideas. Instead, it clearly demonstrates design, purpose, and creation by the omnipotent and omniscient First Cause of all things—the Creator.
5. Sir Fred Hoyle, a world-renowned mathematician and astronomer, attempted to calculate the odds that life could evolve through random processes without engineering and creativity. He concluded that natural forces, acting randomly, would only have a 1 in $10^{40,000}$ chance to produce a simple cell. Since most scientists agree that the total number of atoms in the entire universe is only 10^{80}, the possibility that a cell could come into being by chance is utterly ridiculous.

Have Scientists Created a Synthetic Cell?

Brian Thomas, M.S.

Scientists claim they have successfully created "synthetic life" for the first time. In a 15-year project, a team led by genomics pioneer Craig Venter synthesized DNA from inanimate chemicals. Headlines through several news agencies announced they had created a living cell, but what exactly did these researchers accomplish?

In their study published in *Science*, the researchers of the J. Craig Venter Institute used machines to synthesize DNA, which they then inserted into already living cells. However, the particular DNA sequence they manufactured was an exact copy—except for precisely placed "watermark" alterations—of pre-existing DNA from a living strain of bacteria that had been selected for its ability to be cloned and reinserted into a bacterium.

The researchers went through plenty of bacterial strains to find one with DNA that could undergo the transfer and cloning processes. After that, it was a matter of artificially synthesizing DNA to exactly match the strain that they knew would work. Even after this, it didn't initially succeed. An error had crept into their synthesized DNA, and "success was thwarted for many weeks by a single base pair deletion in the essential gene *dnaA*."[1]

This illustrates the high level of specification that had been built into these bacterial genomes in the first place.[2] In order for their endeavor to succeed, the researchers had to conform their DNA sequence, in all the critical places, to that of the bacteria's. In their words, "this project was critically dependent on the accuracy of these [original bacterial] sequences."[1]

DNA is a very long two-stranded molecule composed of four repeating chemicals, like beads on a string, called bases. The base on one strand pairs with a particular base on the opposite strand, forming a base pair. Many genomes have millions or billions of base pairs, but the genome of the tiny bacteria that these researchers copied was only 582,970 DNA base pairs long.

They sequenced every base, transferring the data to a computer. They then synthesized new DNA to precisely match the sequence. Due to limitations of the DNA synthesizer, they had to start by manufacturing over 1,000 individual lengths, each with approximately 1,080 base pairs. This included extra DNA required for splicing the lengths together. The synthesized genome was then transferred to yeast, which can accurately copy long sequences of DNA and have enzymes that maintain DNA integrity. Finally, the researchers transferred the laboratory-synthesized, yeast-cloned DNA into a living bacterium that had its own DNA removed. The resulting cells grew and multiplied successfully in the lab.

So, after millions of dollars and untold man-hours, pre-existing information was copied from cells into computers, and then placed back into living cells by purposefully manipulating both man-made and cellular machine systems. The resulting cell was therefore not wholly synthetic—only its DNA. But other than four added watermark sequences that served to verify the results, even that DNA was an exact copy of an already functioning bacterial genome.

Despite the headlines, the scientists did not create a bacterial cell from scratch. Instead, they "refer to such a cell controlled by a genome assembled from chemically synthesized pieces of DNA as a 'synthetic cell', even though the cytoplasm of the recipient cell is not synthetic."[1] But the cytoplasm has the machines required for all necessary cellular tasks like carrying sugars, copying DNA, taking out trash, converting energy, regulating production speeds, manufacturing proteins, communicating with the environment, and so on. None of that was artificially synthesized.

In what may be an attempt to add gravitas to this research, Venter told the *Financial Times*, "We have passed through a critical psychological barrier."[3] Surely, this is a reference to ideas about the creation of life being the sole domain of God. But there are reasons why it would be an overstatement to say that this synthesized DNA represents some kind of "divine" accomplishment.

First, there is no biblical mandate that precludes mankind from attempting to build bacteria. Second, since bacteria do not breathe, they do not possess the "breath of life" that the Creator built into certain animals. So, like plants, bacteria do not have a soul or "life principle" and can be considered just very highly organized matter.[4]

Thus, even if scientists can eventually create an entire self-replicating cell—including every working part—from scratch, they still will not have "passed through a critical psychological barrier," because they will only have succeeded in adding fantastic amounts of organization to previously existing matter. Such an organism would not have "life" in the same sense that humans do. Souls are not matter, yet they mysteriously reside in certain creatures.

Overall, this research could serve at least two good purposes. The biotechniques that these scientists pioneered could improve medical technology. Also, by encountering the specificity with which these bacterial cells are constructed, investigators can get a closer appreciation for the engineering genius of their real Architect. In light of what the Lord Jesus accomplished in creating whole, reproducing cells without a reference template, what little these researchers achieved nevertheless "was complicated and required many quality control steps."[1] How much more control was therefore required to have invented the whole cell in the first place, and how much more plain can the evidence for a Creator be?[5]

If anything, this research verifies that His handiwork is wondrous. If a team of brilliant scientists only succeeded in copying information from a germ to a computer and back to a germ, then the Originator of that information must be far more brilliant.

References

1. Gibson, D. G. et al. 2010. Creation of a Bacterial Cell Controlled by a Chemically Synthesized Genome. *Science*. 329 (5987): 52-56.
2. See also Thomas, B. Bacteria Study Shoots Down 'Simple Cell' Assumptions. *ICR News*. Posted on icr.org January 4, 2010, accessed May 25, 2010.
3. Cookson, C. Scientists create a living organism. *Financial Times*. Posted on ft.com May 20, 2010, accessed May 20, 2010.
4. Criswell, D. C. 2009. *Origin of Life*. Dallas, TX: Institute for Creation Research, 13-15. See also Morris, J. 1991. Are Plants Alive? *Acts & Facts*. 20: (9).
5. Guliuzza, R. J. 2010. Natural Selection Is Not "Nature's Design Process." *Acts & Facts*. 39 (4): 10-12.

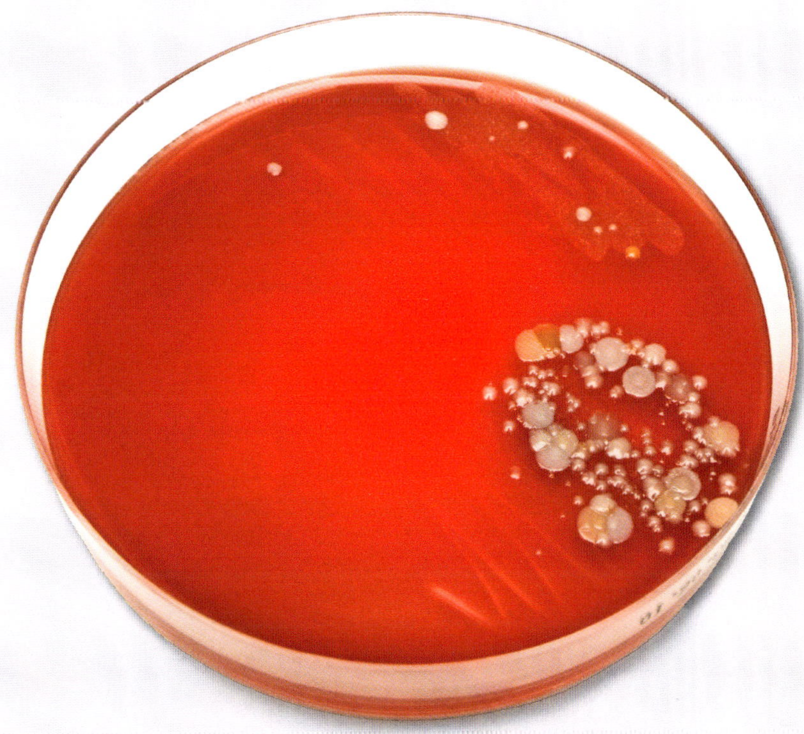

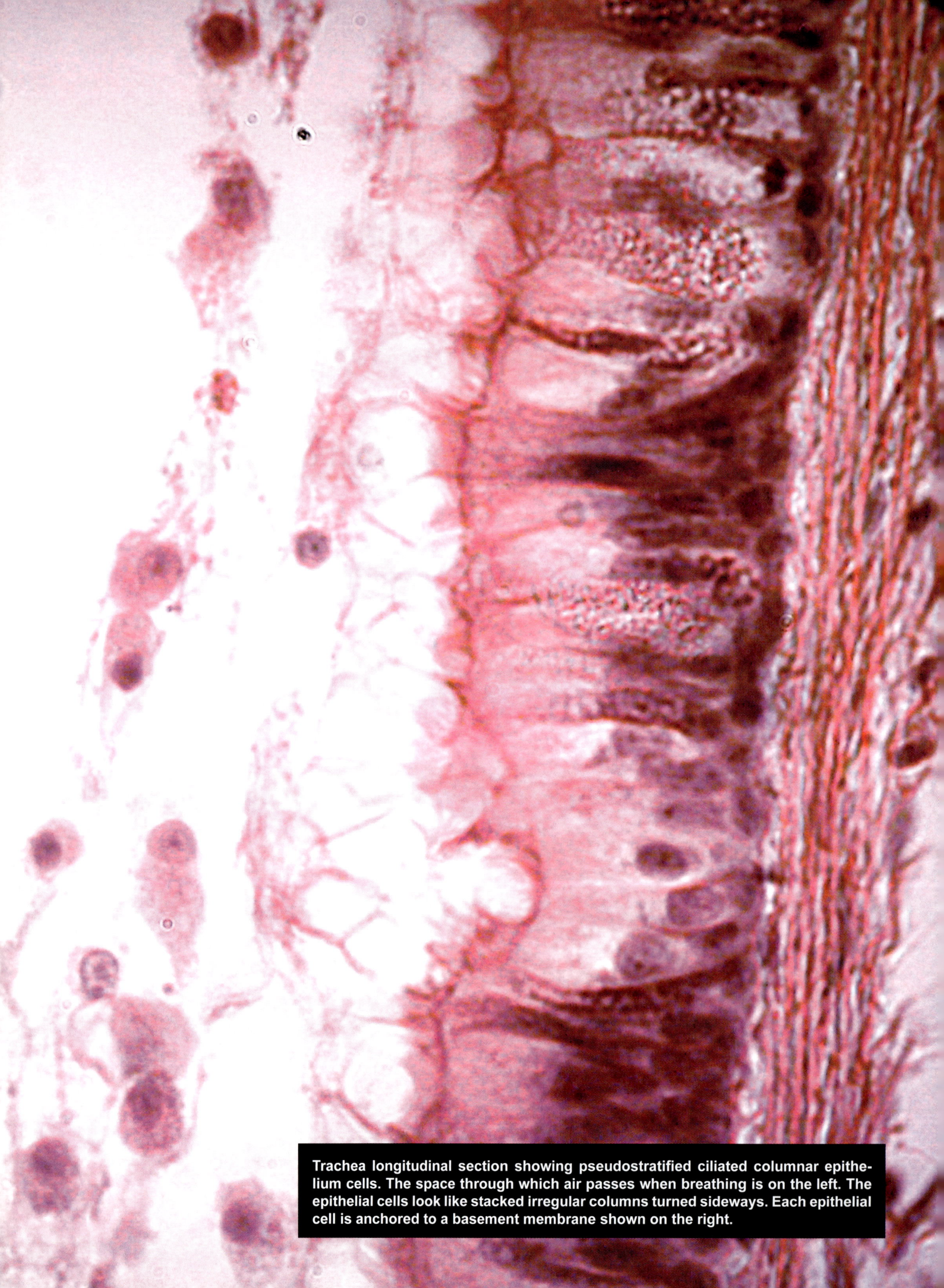

Trachea longitudinal section showing pseudostratified ciliated columnar epithelium cells. The space through which air passes when breathing is on the left. The epithelial cells look like stacked irregular columns turned sideways. Each epithelial cell is anchored to a basement membrane shown on the right.

Chapter 2

A Basic Description of Cells and Cell Types

A Basic Definition of Cells

In 1665, before much was known about cells, a prominent English scientist named Robert Hook conducted research in the early development of telescopes and microscopes. While using one of his microscopes to study plant tissue from the cork oak tree, Hook noticed the cellular structures contained in cork tissue, which is the layer of tissue located just underneath the bark of the tree. Hook described the compartmentalized structures he observed as having the appearance of small rooms and coined the term "cells" (Figure 2.1).

The word "cell" comes from the Latin root *cellula,* which means a small room. In a sense, that's exactly what cells are—small biological compartments with their own self-contained system that is protected and separated from other cells and from the outside world. On the other hand, just like rooms in a house are not completely independent—having connections to other rooms and perhaps even to the outside environment—cells in a complex organism are also interconnected. In the case of multicellular organisms, cells are the basic building blocks for organs, tissues, and ultimately the whole organism.

In a multicellular system, cells are designed to be independent. However, cells are also designed to func-

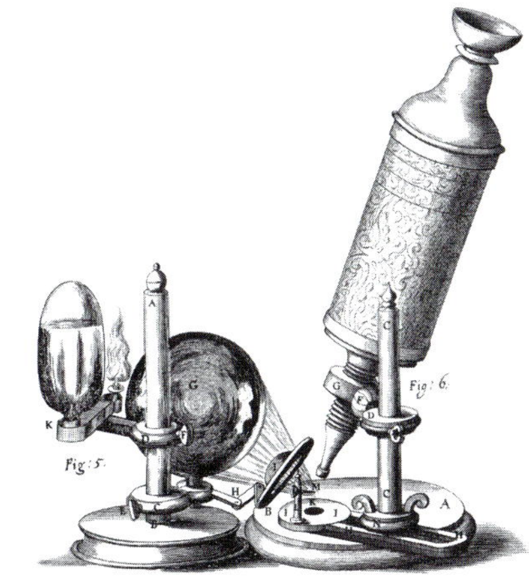

Figure 2.1 This is a drawing from Robert Hook's book *Micrographia* showing how the cellular structure of cork tissue looked like small rooms, which led to the term "cell." In the right panel is a drawing of the microscope Hook used to study cell structure.

tion as an integrated and interactive group. Although organs in the body are made up of groups of individual cells, these groups of similar cell types function together to perform many complex tasks and to respond in very specific and coordinated ways to a wide variety of stimuli from other parts of the body and from the environment. The complexity of cell function and cell response occurs through elaborate systems of communication within the cell itself, between neighboring cells, and with cells in other parts of the organism. These interconnected systems are often referred to as biological networks.

Figure 2.2 A DNA sequencing lab using high-throughput robotics at a government-funded facility.

In the earlier days of cell biology research, scientists were only able to study one gene or one protein at a time. With the advent of high-throughput robotics-based lab technologies, scientists are now able to study the activity of thousands of genes and proteins in a single experiment (Figure 2.2), which allows evaluation of a wide variety of cellular networks and pathways related to diverse biological processes. In fact, the more we learn using these technologies, the more we realize how complicated and intricately engineered life really is. In other words, it is becoming absolutely clear that the complex nature of cell biology did not come about through chance random evolutionary processes.

A term that is often referenced to illustrate this concept is *irreducible complexity*, which refers to the cell being a very complex, highly designed biological machine in which each of the thousands of parts must have been engineered and fully integrated for it to be functional. The evolutionary model cannot adequately explain or account for the immense internetworked complexities that are being revealed through modern cell biology research.

Before discussing the major categories of life defined by their cell structure and the major differences between cellular systems, it is important to highlight some basic features and attributes that are common to all types of cells.

Working from the outside in, the first feature at the cell perimeter is the *plasma membrane*. The plasma membrane is the outer selective barrier that is formed of individual phospholipid molecules that have a hydrophobic tail (repels water) and a hydrophilic head (affiliates with water). It is called a phospholipid bilayer because the phospholipid molecules form a double layer with their hydrophobic tails pointing inward toward each other (Figure 2.3). These plasma membranes are actually quite complex and dynamic, functioning as more than just a barrier between the inside and outside of the cell. Some key functions of the plasma membrane involve the import and export of chemical compounds through specialized transmembrane channels, sensory and signaling processes via specialized receptor proteins imbedded in the membrane, and osmotic (water) regulation of the cell through specialized portals. Contrary to the evolutionary idea that the plasma membrane is a derivative of some primitive, simple, self-assembling lipid structure, the plasma membrane in real cell systems is an extremely complex apparatus. In fact, the development of a functional plasma membrane is a complicated process that involves advanced forms of lipid molecules called *sphingolipids*.

Within the plasma membrane is the internal cell matrix, often called the *cytosol* or *cytoplasm*, which is a semi-fluid substance where all of the inner components of the cell are located. The cytoplasm provides the basic cell matrix for all of the cell's structures and internal ac-

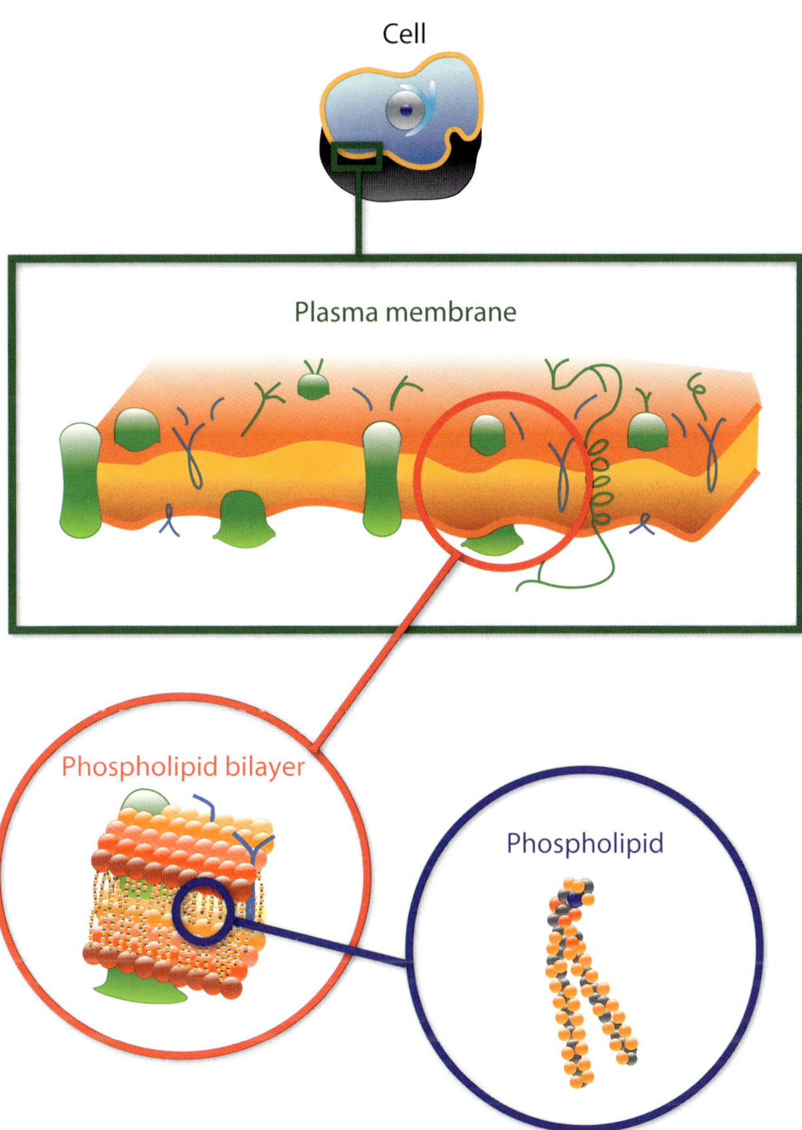

Figure 2.3 An interconnected diagram showing the location of the plasma membrane and the general overall makeup of the phospholipid bilayer membrane structure, along with imbedded macromolecules such as various types of proteins and carbohydrates. A wide variety of proteins provide transport channels across the membrane, cell signaling/communication, and places of support for the cell cytoskeleton—the matrix support structure within the cell. A close-up view of the phospholipid bilayer shows the polarity of the phospholipids, with hydrophobic tails pointing inward and the hydrophilic heads pointing outward. This type of orientation effectively produces a watertight seal between the inside of the cell and the outside, allowing the cell to closely regulate its chemistry. A picture of the phospholipid molecule shows the forked tail structure of the hydrophobic end and the more chemically complex single-head nature of the hydrophilic end.

tivities to occur. Like the plasma membrane, the complexity of the cytoplasm seems to grow with every new discovery in cell biology.

Two Basic Cell Types in Biology

There are two basic cell types in biology, *prokaryotes* and *eukaryotes*, which are clearly demarcated by their different cellular structures and biochemistry. Prokaryotes are single-cell microbes and consist of bacteria and archaea (Figure 2.4). While they share many commonalities, scientists believed that there were enough significant differences between bacteria and archaea to separate them into different domains. Although it is beyond the scope of this book to describe all the details of the archaeal domain, these prokaryotic microbes are a fascinating part of God's creation and occupy some of the most extreme and inhospitable habitats on earth, such as environments high in acids or salts, and also locations of extreme heat. Of course, there are also various bacteria capable of living in these hostile environments. These types of extreme microbes are often referred to as *extremophiles*. Their existence provides a strong testament to divine creation because they occupy environments that are hostile and prohibitive to any sort of so-called evolutionary process. Even though a distinction can be made between bacteria and archaea, they share key distinguishing characteristics that make them both prokaryotes. For instance, they both lack cellular organelles (e.g., nucleus and mitochondria) and both are small in physical size.

Prokaryotes do not occur in the form of multicellular organisms, but exist as free-living single cells. Many prokaryote genera, however, do form large masses of interactive colonies and exhibit elaborate communication mechanisms between and among cells. Masses or mats of bacteria that often occur at the interface between one environment and another are termed *biofilms* (Figure 2.5). An example would be the formation of a biofilm on some sort of solid surface such as a rock in a pool of water. The bacteria in the biofilm would derive the benefit of a support structure and perhaps minerals from the rock environment, and, of course, water and other substances from the surrounding pond environment. Often these colonies include some type of sticky or slimy excreted matrix for the bacteria to live in and that also enables the bacteria to attach to some

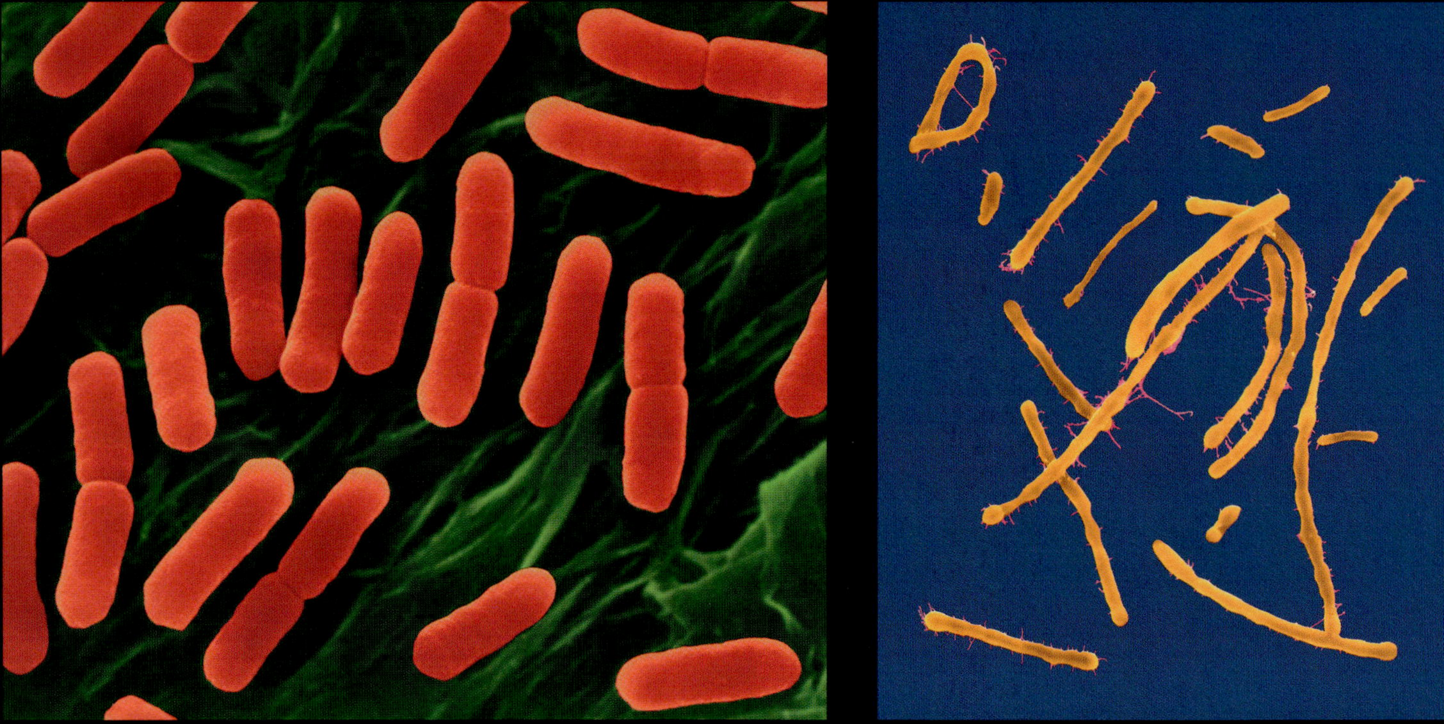

Figure 2.4 Electron micrograph images of prokaryotes from the two domains, bacteria and archaea. *Escherichia coli* bacteria are pictured on the left. The extremophile archaeon *Halobacterium* sp. strain NRC-1, a high salt-tolerant species, is shown on the right. Not all archaea are extremophile organisms, and there are also many bacteria that are classified as extremophile.

Figure 2.5 Thermophile microbes clump together and form colorful large mats and films at the Grand Prismatic Spring at Yellowstone National Park.

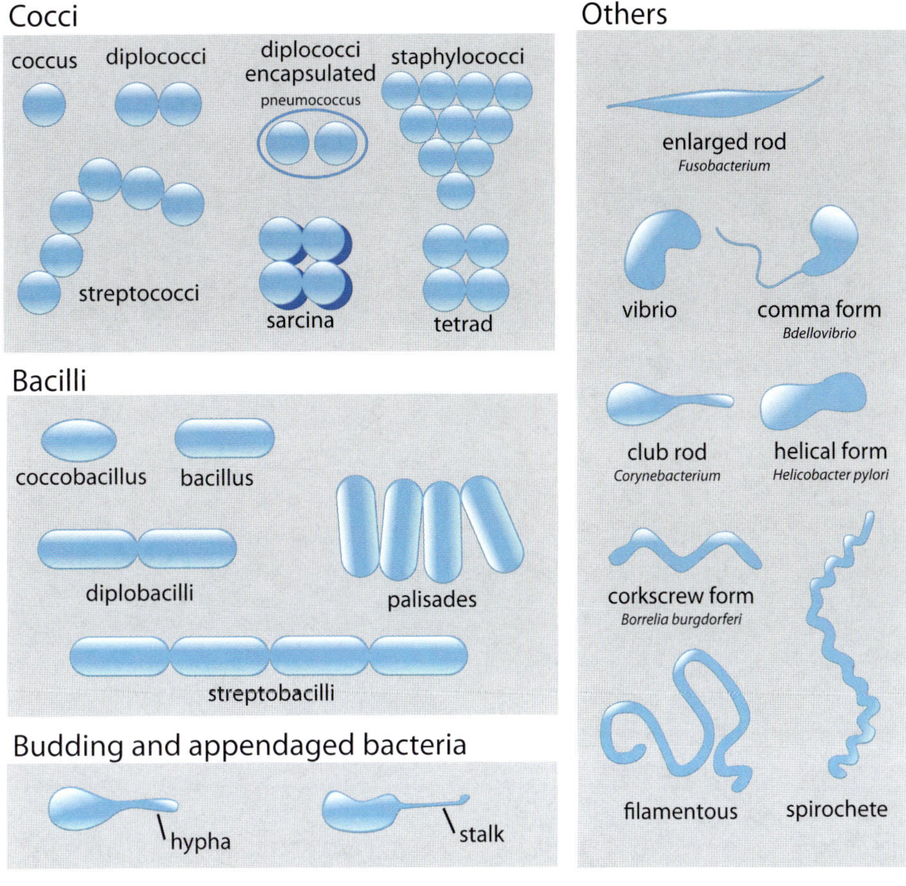

Figure 2.6 Bacteria come in different shapes, and in some cases form distinctive colony structures and configurations in association with other bacteria of the same kind.

sort of substrate like a rock. The prokaryotes that we are most biologically familiar with from a research perspective are those associated with human health, particularly those of a pathogenic nature (i.e., those that cause disease).

Eukaryotes include both single-cell organisms and multicellular organisms. Eukaryotes have key cellular features that are truly distinctive compared to prokaryotes—a nucleus and organelles (e.g., mitochondria and, in the case of plants, chloroplasts). A nucleus is a membrane-bound organelle within the cell that contains the main chromosomal DNA. As mentioned earlier, prokaryotes lack a nucleus. The DNA of bacteria is located directly within the cell's cytoplasm and is not enclosed by a membrane. The structure and function of the various cell organelles in eukaryotes will be discussed later.

One aspect that is evident when comparing cell types is that the level of complexity is typically greater in eukaryotic cells than in prokaryotes. In addition, the presence of certain intracellular structures in eukaryotes, along with the added complexity, results in differences in cell size. Generally speaking, eukaryotic cells (10–100 µm in diameter) are typically larger than prokaryotic cells (1–10 µm in diameter). There are many other features and variations in cell structure that distinguish the differences between prokaryotes and eukaryotes. These will be discussed as we focus on each cell type.

Prokaryotic Cell Characteristics

For the purposes of this text, we will discuss prokaryotic cell structure and biology strictly within the bacterial domain. Figure 2.6 shows the basic types and shapes of bacterial cell structures found among prokaryotes. Bacteria can be studied and characterized by their shape and whether or not they form colonies or some other social configuration with other bacteria of the same kind.

In general, there are four major groupings of bacteria based on cell shape and the tendency to form multicell groups: cocci (round cells); bacilli (oval or rod-shape);

budding and appendaged bacteria (elongated shapes, new cells formed by branching and budding); and then all others. The bacterial kinds in the "all others" group assume a wide variety of shapes, including kidney beans, bananas, corkscrews, and snake-like shapes. The physical appearance of a bacterial cell is often referred to as *cell morphology*. The morphology of cells can be observed by staining the cells and observing them under a microscope. The various features of the cell can be stained to aid in visual observation. This differential staining is accomplished through a specific cell-staining chemical.

Even though bacteria come in a vast array of shapes, they do share a number of structural features in common. Figure 2.7 provides a visual guide to the major features of bacterial cell structure. Beginning from the outside and working inward, the first structure encountered in bacteria is the *capsule*. The capsule is a polysaccharide matrix containing some water and serves a variety of functions, including protection from being engulfed by other cells, defense against viruses, cell desiccation control, and protection from toxic hydrophobic compounds. It also provides a means for the bacterium to attach itself to some solid substrate or to other bacteria. Because the capsule is water soluble, it is difficult to stain compared to the rest of the bacterium. It often appears as a clear ring around the outside of the bacterium when observed under a microscope.

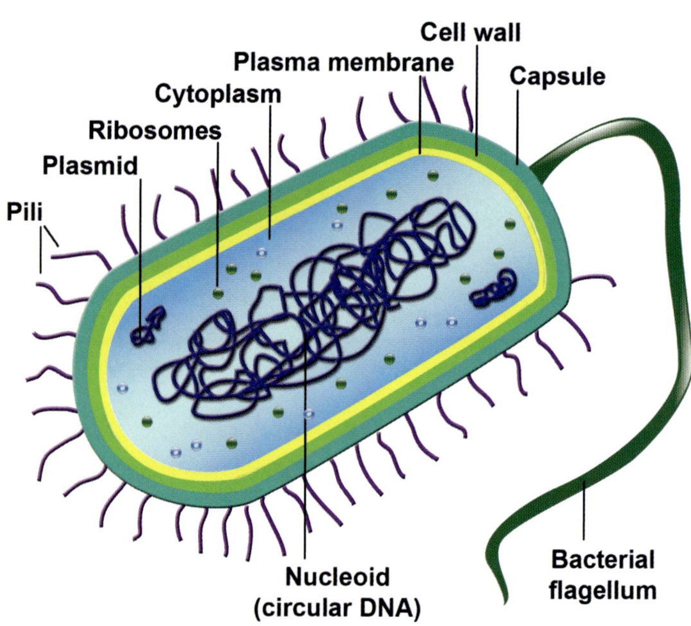

Figure 2.7 A model of a prokaryote cell showing the primary structural features common to many bacteria.

Another major outside structure or appendage that is highly visible is the *flagellum*, a tail-like structure that can spin and propel the bacterium around its environment. The flagellum is often used as an example of a cell-based design feature that confounds evolution. The whip-like appendage is imbedded in the outer shell of the bacterium and powered by a miniature motor with many of the same design features found in man-made electric motors, except that it is powered by protons instead of electrons (Figure 2.8). The flagellum motor is capable of amazing action that cannot even be observed in man-made motors, such as instant stop, start, and reverse capability; and 100 percent efficiency at cruising speeds. In fact, its top cruising speed is comparable to a 6-foot-tall human swimming at 60 mph.

Just within the outer capsule is the bacterial *cell wall*, which provides the rigidity and shape for the bacterial cell. The cell wall also protects the more inward parts of the bacterium from the outside world. The unique molecular characteristics of the bacterial cell wall have enabled scientists to divide the bacterial domain into two main groups called "gram positive" and "gram negative." Both groups have a cell wall composed largely of a *peptidoglycan* (protein-carbohydrate) matrix. The gram-positive types have a very thick cell wall that traps staining molecules, making them appear darkly stained under a microscope. Gram-negative types have a very thin cell wall that does not stain very well. Therefore, they appear very light or pink-colored under a microscope. The process of gram staining was named after Hans Christian Gram. In 1884, he developed a technique that involves the application of a series of chemical stains and washes. Figure 2.9 shows the gram-positive anthrax bacteria after gram staining.

Located just underneath the cell wall is the *plasma membrane*. The basic structure of the plasma membrane was discussed earlier and illustrated in Figure 2.3. Interestingly, gram-positive bacteria also have a second, very thin membrane outside their cell wall. This is used for more highly refined cell communication, environmental sensing, and signaling.

The inside matrix of the cell is referred to as the *cytoplasm*. In bacteria, the cytoplasm does not contain organelles and is not highly compartmentalized, like it is in eukaryotic cells. There are, however, several distinctive features associated with bacterial cytoplasm. One is the *nucleoid*, the location of the bacteria's main chromosome. The bacterial chromosome is located directly in the cytoplasm, typically toward the middle of the cell, and is not compartmentalized or encased in a membrane. Most bacteria have only one main chromosome and it is usually circular. In addition to the main chromosome, many bacteria also have *plasmids*, small circular pieces of DNA that often carry various types of important genes. Bacteria can actually exchange genetic information with each other by transferring plasmids through tubular connections using structures called *pili* in a process called conjugation. Figure 2.7 shows pili located on the surface of a model bacterial cell. Figure 2.10 shows a diagram of two different strains of *E. coli* exchanging genetic information through conjugation. Scientists have exploited and modified the bacterial plasmid system for use in many molecular biology applications that require the manipulation of DNA fragments. By inserting (splicing) foreign DNA fragments into specially engineered plasmids, they can produce lab strains of *E. coli* that propagate and maintain the foreign DNA fragment. This process is termed "DNA cloning" and is distinct from cloning animals or humans, which we will cover in a later chapter.

Another important component of the bacterial cytoplasm is the *ribosome*, a protein complex that produces (translates) proteins from RNA transcripts, also known as messenger RNA (mRNA). Gene activity and protein production will be discussed in Chapter 4. While both eukaryotes and prokaryotes have ribosomes, the ribosomes in bacteria are generally not physically associated with any major subcellular structures and they can freely attach to an mRNA in the cytoplasm to make proteins. In fact, multiple ribosomes will attach to a single transcript, often while the mRNA is still being transcribed. Under a microscope, the structures formed by ribosomes traveling down an mRNA and producing a protein look like Christmas trees and are often referred to as such (Figure 2.11).

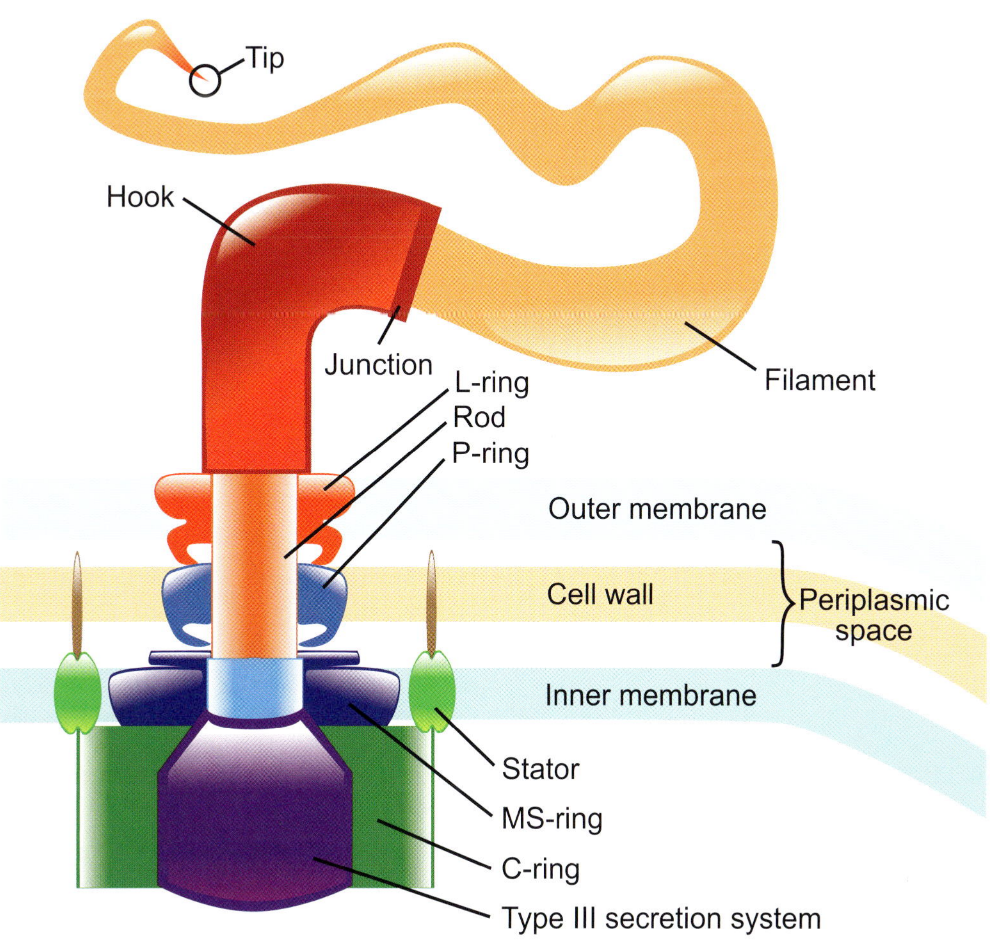

Figure 2.8 A model of the bacterial flagellum and its motor apparatus found in many gram-negative bacteria.

Development of a Model System for Cell Biology

Much of our understanding about cellular systems has been gained through research that is focused or standardized on a specific set of organisms. In the early days of modern cell biology research (during the 1950s), scientists decided to agree on research focused on the *E. coli* bacterium, a natural microbial inhabitant of the human intestinal tract. This served as the first standardized biological "model system."

By selecting a single organism to study, scientists working on different aspects of cell biology could integrate their data because they were based on the same cellular system and overall cell biochemistry. This also provided a convenient way to rapidly develop and expand research projects because they were all based on a resource of foundational and regularly increasing knowledge for a standardized cell system. Quickly and efficiently developing research knowledge in this way gave scientists an advanced body of knowledge that could be applied to the study of other bacteria and even to basic processes in eukaryotic cell systems.

Over the years, scientists began adding a variety of other model prokaryote and eukaryote cell systems. Much of what we know today about basic cell biochemistry was first researched and discovered using the *E. coli* system. In fact, the common laboratory suite of methodologies for cloning, expressing, and maintaining DNA fragments currently is and always has been based on *E. coli*. This model prokaryote is truly the laboratory workhorse of modern molecular and cellular biology.

E. coli: The Prokaryotic Model

One of the top scientific stories in 2009 was the summary of the genetic changes observed in a long-term laboratory experiment involving *E. coli*. This experiment analyzed the genetic changes that occurred through generations of growth in various types of biological media. The study, performed in the lab of Richard Lenski at Michigan State University, began in February 1988. Twelve separate cultures of *E. coli* bacteria were grown in various types of liquid media to provide nutritional challenges related to carbon sources. This experiment was conducted to provide a type of selection pressure that would favor mutations that altered proteins involved in metabolism. The experiment lasted approximately 20 years, resulting in the growth of 40,000 generations, accomplished by sub-culturing the *E. coli* bacteria after every 24 hours of growth, transferring 1 percent of each population to new liquid growth media.

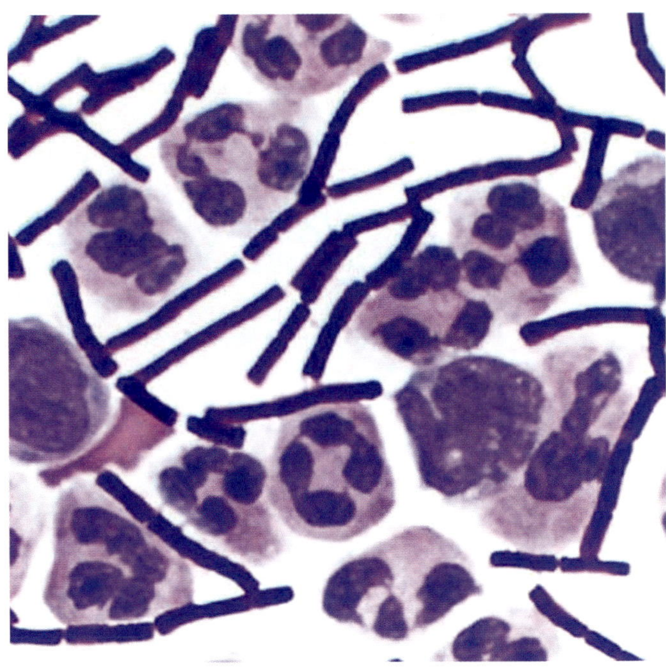

Figure 2.9 The dark purple (gram-positive) rod-shape structures are anthrax bacteria that have been gram stained in a sample of cerebral spinal fluid.

Many research papers were published over the course of the project on various isolated aspects of the study, but the primary "evolution" paper was published in 2009 in the prestigious journal *Nature*.[2] In this milestone research publication, bacteria were sampled at 2,000; 5,000; 10,000; 15,000; 20,000; and 40,000 generations from the *E. coli* population that had been grown under sub-optimal levels of glucose. Upon each sampling, sub-samples of the bacteria were frozen cryogenically, awaiting the development of DNA sequencing technology that could cheaply and quickly sequence the genomes from each generation for comparison. This technology became available in the mid-2000s and the genome of the bacteria from various generations was sequenced. By comparing each genome sequence, the development of mutations in the genome over time could be cataloged and compared to the overall health and vitality (fitness) of the bacteria as determined through physiological measurements.

A critical examination of the famous *E. coli* "evolution"

2 Barrick, J. E. et al. 2009. Genome evolution and adaptation in a long-term experiment with *Escherichia coli*. *Nature*. 461 (7268): 1243- 1247.

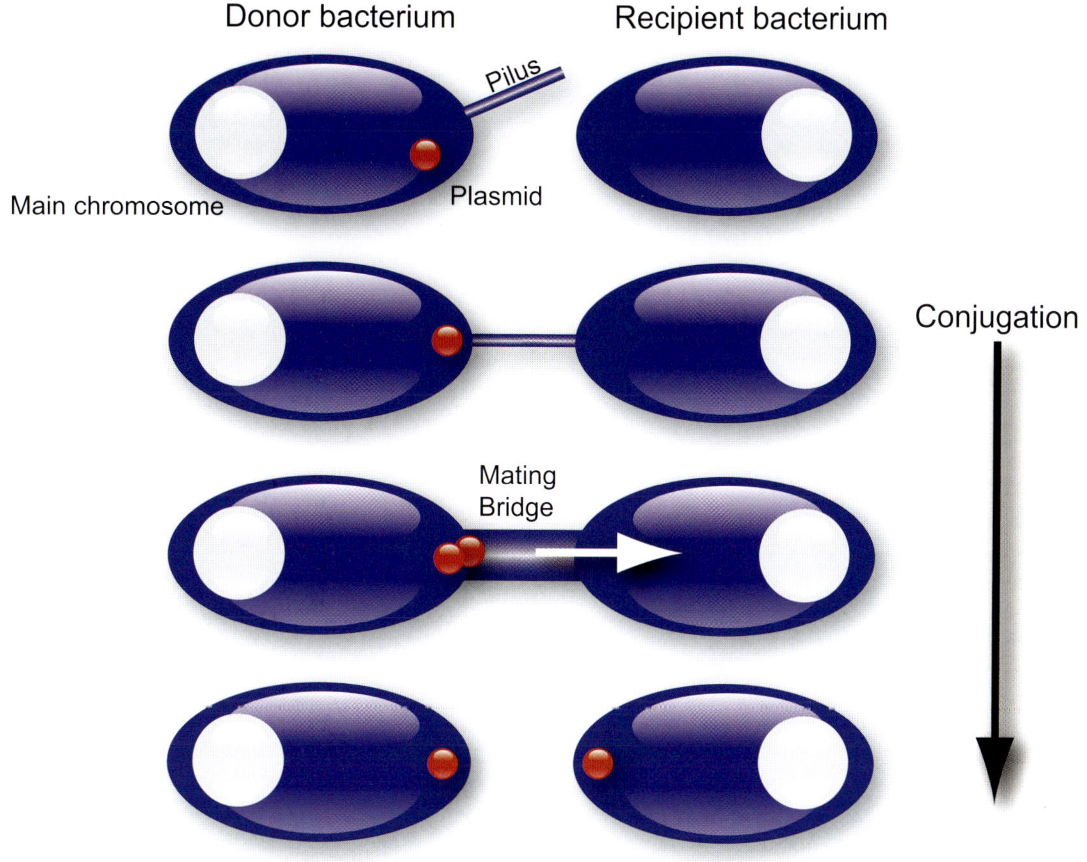

Figure 2.10 An image of two *E. coli* cells in the process of conjugation, exchanging genetic material through a pilus.

Nature paper reveals that the authors did not achieve data favorable to the evolutionary model and that they attempted to spin the results in the most positive light possible, given the circumstances. While the data provide very little evidence to support evolution, the results do favor the creation model. The data show that the bacteria adapted to the change in media immediately at the beginning of the experiment. Remarkably, this happened within the initial set of generations, not later after thousands of generations.

This trend supports other studies performed in higher animals that show that God has placed within each created kind a certain adaptive level of genetic diversity—and possibly other genetic mechanisms—which allows for rapid adaptation to new environments, followed by organism stasis. A created kind does not become a different type of creature, but makes slight changes in its system to adapt to the new environment. The adaptation is based on pre-engineered, built-in survival mechanisms. This is what was observed with the long-term *E. coli* experiment. The bacterial populations made slight adaptive changes in biochemistry at the very beginning of their introduction to a new environment (rapid increase in fitness), and then exhibited stasis after that (little change in fitness). In addition, the accumulation of random mutations over time did not confer an increase in fitness. Oddly, the researchers did not report fitness levels after 20,000 generations (the halfway point in the experiment), but continued to report an increase in mutation rates up to 40,000 generations. Why was the fitness data omitted for the second half of the experiment? The most likely answer is that fitness probably declined rapidly during the second half of the experiment while mutations were accumulating to highly detrimental levels.

The Role of Cells in Biological Creation

Cells are the basic structural building blocks of every life form. In the creation model, they are the foundational unit God used in creating all types of life. As an analogy, consider how a contractor constructs a new building. He essentially uses the same basic structural units for all different types of architectural structures, such as pieces of wood, bricks, sheetrock, etc. He may use different types of brick or wood planks in one structural design as compared to another, but the general features of each type of building block are similar and serve a similar function. We see the same design principles or concept in biological creation. Different types of microbial, plant, and animal system architectures are put together with *cellular* "building blocks." The cell is the basic functional unit of life in everything from microbes to humans. Cells are found in all types of organisms and are as essential to life as atoms are to molecules.

Not only do cells provide structure and support to higher-level animal systems, but cells also provide function as well. For instance, muscle tissue movement not only requires individual activity at the level of the single cell, but it also requires all of the cells in the muscles and nervous system working together to provide the needed action. These processes are mediated by highly complex and clearly designed communities of internet-worked cells. The biochemical complexity present in a single cell, every tissue, and the whole organism is highly complicated and finely tuned.

Cells-based systems of biological architecture display shared commonalities among structures across the spectrum of life. Scientists have misinterpreted this feature in an attempt to fit the repeated similarities in design patterns that are observed in the biological data to a flawed naturalistic worldview. However, these commonalities are not an indication of some common ancestor or evolutionary path, but rather the reuse of optimally effective design patterns by a single designer, God the Creator. In general, the more similar one organism's overall architecture is to another, the more cellular design features will also be shared.

However, each basic category of organism has features unique to its structure and biology that are hard-coded in its DNA. These unique cellular features form the foundation of what we observe visually in the architecture of the whole organism and provide a method to categorize life based on the "created kind principle" described in Genesis. In other words, we recognize a chicken when we see one and know that it is clearly different from a cat or a dog. While there is some variation within created kinds, there are also obvious distinct and observable boundaries between kinds. These boundaries reflect the scientific principle stated on multiple occasions in Genesis—organisms reproduce and genetically perpetuate "after their kind."

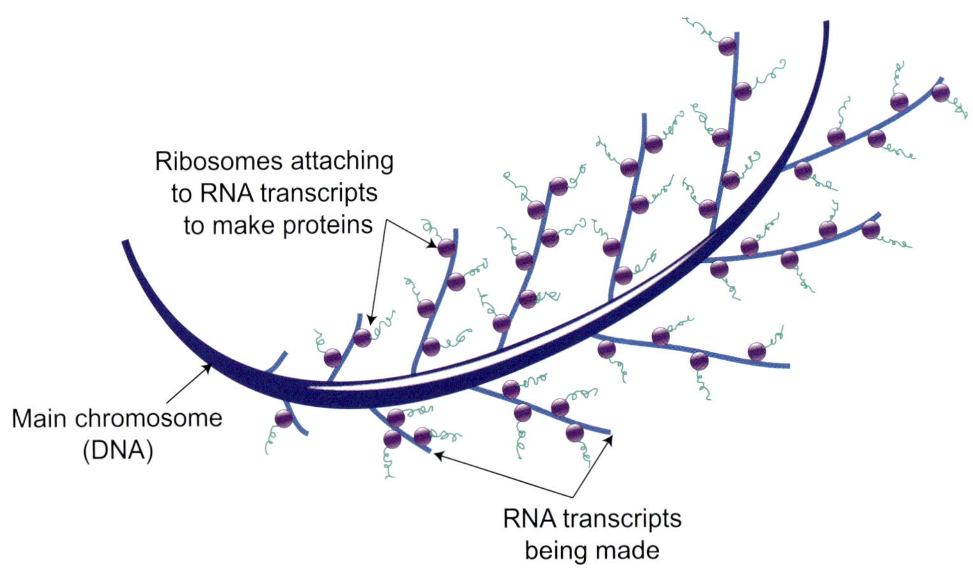

Figure 2.11 Under a microscope, structures that look like Christmas trees are formed by ribosomes traveling down an mRNA and producing a series of branches or developing proteins.

Summary

1. *Irreducible complexity* is a term often applied to the cell, which is a very complex, highly designed biological machine in which each of the thousands of parts must have been engineered and fully integrated for it to be functional. In simple terms, all of its parts must be present and working before the cell can function at all. Parts and pieces that do not work, or are not yet available, would eliminate the possibility of the existence of the cell.
2. A *prokaryote* is a single-cell organism without a distinct nucleus or other specialized internal structures. Bacteria are the most common type of prokaryotes. They come in a vast array of shapes and share a number of common features.
3. Although prokaryotes are single-cell organisms, they are quite complex. The *Escherichia coli* bacterium (*E. coli,* commonly found in the human intestinal tract) has been widely studied. Its highly visible and marvelous *flagellum*—a tail-like structure with instant stop, start, and reverse capability—is used to propel the bacterium around its environment. Evolutionists have a very difficult time explaining how the flagellum and other cell features came into being.
4. A famous 20-year experiment with *E. coli* demonstrated that the bacteria adapted very quickly to new environments. Although the experiment ran to 40,000 generations, the *E. coli* became less viable as genetic mutations accumulated. Apparently, the built-in design of *E. coli* allows for rapid adaptive changes, but *not* for constant change—certainly not change that would transform it into another kind of bacteria.
5. The same basic concepts humans use to design and build things are evident in cells and the highly complex and clearly designed communities of internetworked cells. The reuse of similar components demonstrates optimal design patterns from a single designer. The more similar one organism's architecture is to another's, the more design features will be shared. While there is some variation within each kind of organism, there are also distinct and observable boundaries between the kinds. The observable facts fit very nicely with the biblical model. Those same facts do not fit with any kind of evolutionary model.

Life's Indispensable Microscopic Machines

Randy J. Guliuzza, P.E., M.D.

Humans love machines. Everyone likes how they save time and make jobs easier. If a new machine is invented that helps detect specific diseases and aids in early diagnosis, it is big news.

People also enjoyed watching machines operate before their many moving parts were covered up for safety reasons. Now, TV programs that slice machines apart so people can look inside are popular. So it is likely that someone would listen intently if a Christian were to describe some real microscopic machines right inside *their own bodies* working to keep them alive.

Be assured that helping anybody learn of these intricate, minute machines will be a powerful testimony of the Lord Jesus' "invisible" qualities, like His endless power and intellect, that are clearly seen by the things He has made (Romans 1:20.)

Cellular Life Operates with Machines

Begin by showing that cellular machines, like man-made machines, consist of numerous interconnected moving parts that function together for an intended purpose—but are far superior. Functioning in repetitive mechanical cycles, cellular machines have chemical molecule "parts" that usually switch between two different—but still very precise—shapes in a strictly controlled manner. Like any motor, they convert fuel into kinetic energy to *make things move* in specific directions. A few examples are:

- DNA maintenance robots that proofread information, unwind the double helix, cut out defects, splice in corrections, and rewind the strands
- Intracellular elevators
- Mobile brace-builders that construct distinct internal tubular supports
- Spinning generators that move molecules from low to high energy states
- Ratchet devices that convert random molecular forces to linear motion
- Motors that whirl hair-like structures like an outboard motor
- A microscopic railroad with engines and tracks

Finally, describe *how* the machines are made to reinforce awareness of the total design process. Ask your acquaintance to visualize an assembly plant that is so advanced and so small that dozens could fit on the head of a pin. The energy to run both the assembly plants and machines is finely tunable and supplied by sunlight, molecular motions, heat, electricity, or chemical conversions.

Machine parts themselves are complex molecules corresponding to switches, batteries, motors, brakes, shafts, rods, hooks, bearings, bushings, springs, end caps, valves, seals, plugs, rivets, spot-welded joints, mounts, and braces. These are fitted together by other molecules that act like templates, work benches, clamps, and vises. Yet other molecules take the final products where they are trimmed, folded, and set to be activated. Another molecular work station will package, label, and transport products to their correct destinations. After the machine's useful life is ended, another apparatus will engulf it, break it down, and send the components back for recycling.

Molecular Machines Are Best Explained by Design

The microscopic size of these machines is vital to systematically fine-tune dozens of molecular properties. For example, many molecular motors must work cooperatively to transform molecular movements to visually detectable levels. All of these features are detailed in advance and the information is stored in the DNA's plans and specifications.

In a typical kitchen, there are food processors, blenders, and mixers. In some ways they *look* similar, but their capabilities actually have little overlap. Each machine has a *primary* function which it performs well based on the speed and manner in which its parts interact. These parts fit tightly together, which means that though a few blender and mixer parts have exactly the same function, one cannot just swap them. In similar manner, the parts of molecular machines are meticulously fitted together for their primary purpose. Many function in totally unique roles that are critical for life. And some cannot lose a single part or exchange parts with other machines without that machine—and the entire organism—breaking down.

In most people's minds, the words "machine" and "designed" belong together, so just knowing that complex cellular machines exist is sufficient evidence to make that connection. But others are stuck in evolutionary explanations for the origination of molecular machines. It would be beneficial to know why this thinking is better at rationalizing than explaining.

Evolutionists Submit Implausible Explanations

When looking at the evolutionist's best scientific journals for explanations specifically on the origins of molecular machines, stay alert for extraordinary extrapolations.[1] For instance, if a window fan is the machine under investigation in one of these papers, be prepared to look for this predictable pattern:

> **Finding:** Researchers discover that fans have electric motors that spin blades to move air.
>
> **Conclusion:** Air conditioners are simple derivations of fans because they also have electric motors that spin blades—notwithstanding an air conditioner's unique motor, blades, compressor, condenser, evaporator, and thermostat.

The exaggeration is assured. Why? Since researchers *find* only one fully functioning machine or another, evolutionary *conclusions* of how, in the remote past, parts from one molecular machine morphed into another will always be conjectures inferred greatly beyond what the findings support.

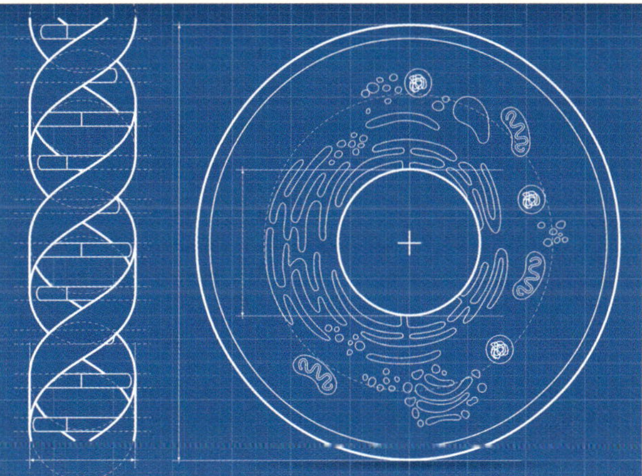

These papers survive peer review still containing extraordinary extrapolations that sidestep real explanations for the arrival of fundamentally distinct parts, instead depicting molecular parts as (somehow) having been stripped from primary functions elsewhere in the cell and spontaneously re-assembled into new machines. Critics point out that those great gaps of information make the evolutionary path unbelievable. Note how these weaknesses are merely dismissed when a top evolutionary authority like Dr. Jerry Coyne says, "It is not valid, however, to assume that, because one man cannot imagine such pathways, they could not have existed."[2] Be content in recognizing that those types of responses cannot be satisfied with scientific answers.

Lacking experimental evidence supporting their explanations about molecular machines, evolutionists have turned to a firm belief that if it can be imagined, it could happen. In conversation, highlight this disconnect. Point out that, given the extreme precision of these machines, evolutionary accounts must repeatedly use three words not normally associated with precision: "recruited,"[3] "cobbled,"[4] and "tinkered."[5]

Thus, evolutionists believe in a simplistic scenario where "the necessary pieces for one particular cellular machine... were lying around long ago. It was simply a matter of time before they came together into a more complex entity,"[6] upon which natural selection tinkered away at cobbling together borrowed parts for millions of years. Aside from the magical whimsy, this explanation is like saying cars originated when an engine was coupled to a transmission, which was mounted to a chassis, and so forth. Leaving another major unanswered question—where did the engine, transmission, and chassis come from?

Pulling It All Together

The best way to appreciate machines is to watch them. Unfortunately, pictures of molecular machines are rare and drawings most likely will not be available during spontaneous conversations. But using words to build mental pictures of these incredible miniature machines can be effective. Human minds powerfully connect machines to design. People know that while some animals may use tools, only humans build machines.

For that reason, a conversation could be very engaging since it may be the first disclosure to most people of the existence of these machines. Why? Because in evolution-based education, not all scientific findings are equally welcome and, thus, are subject to being selectively promoted.

Revealing the convoluted thinking that does attribute precise microscopic machines to blind tinkering—but not to design—will let it be seen for what it is. So, go ahead and tell someone about these life-sustaining little machines. They may build the bridge for someone to find eternal life in their life-giving Creator.

References

1. See Clements, A. et al. 2009. The reducible complexity of a mitochondrial molecular machine. *Proceedings of the National Academy of Sciences*. 106 (37): 15791-15795.
2. Coyne, J. A. 1996. God in the details. *Nature*. 383 (6597): 227-8.
3. McLennan, D. 2008. The Concept of Co-option: Why Evolution Often Looks Miraculous. *Evolution: Education and Outreach*. 1 (3): 247-258.
4. Hersh, B. and S. B. Carroll. 2005. Direct regulation of knot gene expression by Ultrabithorax and the evolution of cisregulatory elements in *Drosophila*. *Development*. 132: 1567-1577.
5. Clements, 15793. Also Jacob, F. 1977. Evolution and tinkering. *Science*. 196 (4295): 1161-1166.
6. Report on Clements' PNAS article (reference 1). Keim, B. More 'Evidence' of Intelligent Design Shot Down by Science. *Wired Science*. Posted on wired.com August 27, 2009, accessed June 5, 2010.

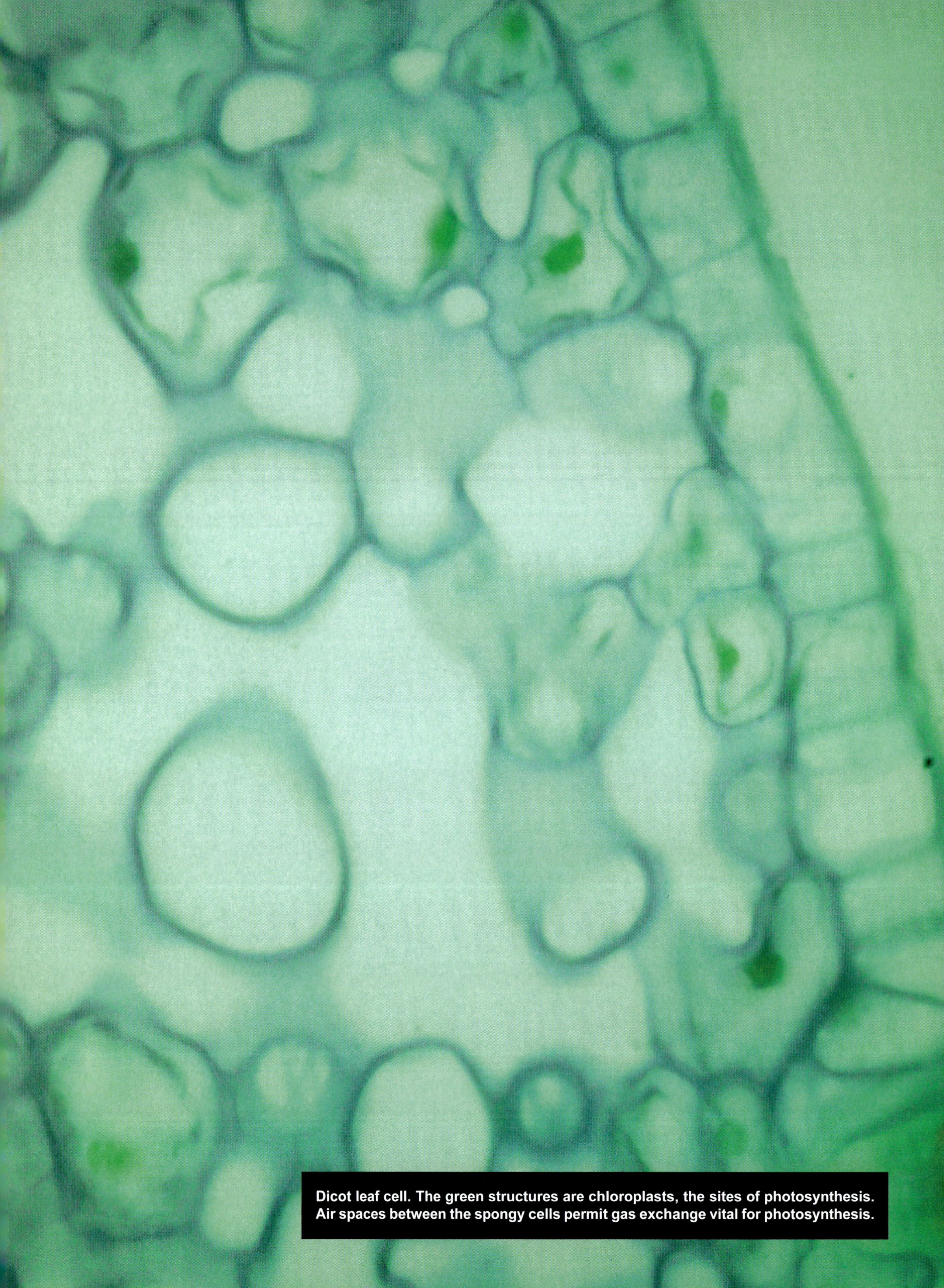

Dicot leaf cell. The green structures are chloroplasts, the sites of photosynthesis. Air spaces between the spongy cells permit gas exchange vital for photosynthesis.

Chapter 3

Eukaryotic Cells and Multicellularity

Eukaryotic cells are more complicated than prokaryotic cells, as exhibited by their increased complexity in cell structure, organization, and function. However, many of the basic biochemical processes discovered in *Escherichia coli* (the prokaryotic cell model) provide an excellent starting point for research aimed at understanding similar, but more advanced, processes in eukaryotic systems, such as DNA replication, RNA transcription, protein translation, and cell metabolism.

Like prokaryotes, some types of eukaryotes can also be found as single-cell organisms (e.g., yeast or amoebas). In fact, the common baker's yeast (*Saccharomyces cerevisiae*) is a single-cell eukaryote. Like *E. coli*, it has been designated and developed as a model biological system that continues to be heavily studied. There are other multicellular model cell systems for eukaryotes that represent different levels of organism complexity and type. Such organisms are the nematode (*Caenorhabditis elegans*), fruit fly (*Drosophila melanogaster*), mouse (*Mus musculus*), thale cress (*Arabidopsis thaliana*), rice (*Oryza sativa*), and human (*Homo sapiens*). (See Figure 3.1.) Even though humans violate almost every qualification required of a model organism (see below), the

Figure 3.1 Pictures of some of the major eukaryotic model organisms. Clockwise from upper left: mouse (*Mus musculus*), thale cress (*Arabidopsis thaliana*), yeast (*Saccharomyces cerevisiae*), fruit fly (*Drosophila melanogaster*), and nematode (*Caenorhabditis elegans*).

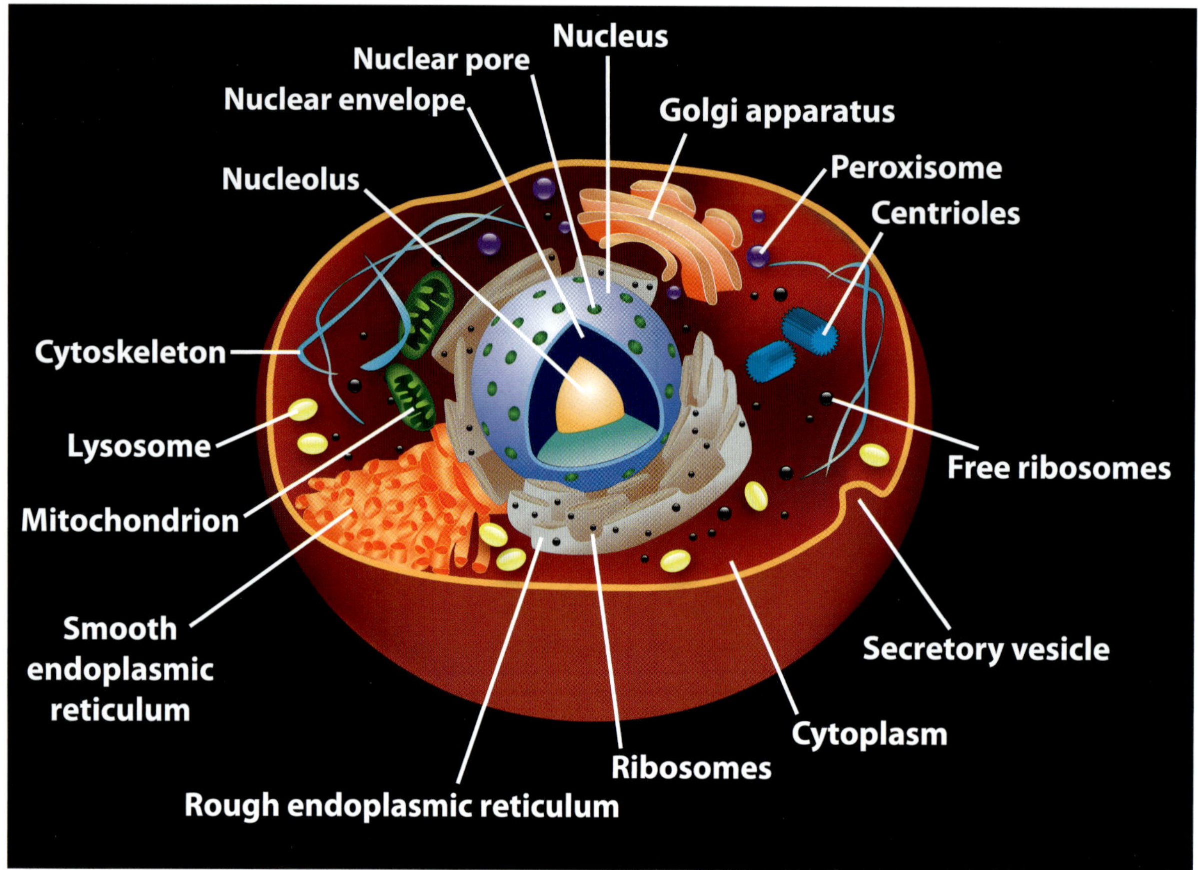

Figure 3.2 Picture of an animal cell containing many of the key features found in a eukaryote.

human system is one of the most vigorously studied because it is advantageous for reasons of medical and health-based research.

In addition to extensive biological and biochemical study of the major model organisms, their DNA has been heavily sequenced, and much of it is publicly available free of charge in a wide variety of searchable online databases.

When decisions were made in the biological research community to select certain higher-level (multicellular) eukaryotes for development as model systems, a number of prerequisites were required:

1. They had to be easily manipulated and reared in the lab.
2. They had to be easy to breed, with short generation times to facilitate genetic studies.
3. They had to have a small number of chromosomes, plus low nuclear DNA content to facilitate DNA sequencing and analysis.
4. They had to preferably be a well-studied system from a basic biology and biochemistry standpoint.

For the purpose of our consideration of eukaryotes, we will discuss cells in the context of a multicellular organism and, in most cases, human cells.

The Highly Engineered Eukaryotic Nucleus and Mitochondria

As mentioned previously, the eukaryotic cell has a number of obvious features and structures that distinguish it from prokaryotes. Figure 3.2 illustrates most of the primary features that are found in a typical eukaryotic animal cell. There are specific subcellular structures that are enclosed by membranes, as well as other cell features that increase the overall complexity of the cell. One of the key differences between eukaryotic and prokaryotic cells is that eukaryotes contain a variety of organelles, which are rather complex membrane-enclosed structures within the cell that allow for the compartmentalization of many important cell processes.

The chief defining subcellular organelle structure in a eukaryotic cell is the *nucleus*. The nucleus contains the DNA or chromosomes and is the primary site for gene regulation and transcription (see Chapter 4). The DNA, which is the chromosomal material in the nucleus, is often referred to collectively as *chromatin*. The regions of chromosomes that contain higher numbers of protein-coding genes are often referred to as *euchromatin*. The regions that tend be relatively void of protein-coding genes and contain various types of repeated sequence are referred to as *heterochromatin*. In most animals and some plants, the chromosomes are divided into sex chromosomes and autosomes (non-sex chromosomes).

Humans have two sex chromosomes, X and Y. Human males have an XY sex chromosome complement, while females have an XX complement. The Y chromosome carries genes that cause the male condition to develop in the embryo. In addition to the set of sex chromosomes, there are 22 sets of autosomes in humans, with each set being numbered 1 to 22 (Figure 3.3).

Some of the key structural features of the nucleus include the nuclear pores, envelope, chromatin, and the nucleolus. The membrane of the nucleus is referred to as the nuclear envelope and contains specialized pores that allow the flow of molecular information in and out of the nucleus. It provides a highly specialized environment for the cell's chromosomes to be maintained, regulated, and accessed. In a sense, the nucleus is the command center for the cell, as it holds the cell's informational database and sends out encoded orders (mRNA gene transcripts) in response to different types of cell signals from a variety of sources.

A prominent feature of the nucleus is the *nucleolus*. It can often be viewed under a microscope, appearing as a dense mass or dark blob within the structure of the nucleus. Scientists have discovered that this is a very specialized region of the DNA that contains primarily ribosomal RNA genes. These are genes that code for the protein subunits that make up the ribosome complex. Because the cell must contain thousands of ribosomes to keep up with protein production, these genes are found in multiple copies.

As scientific study progresses, it is becoming apparent that the nucleus is a masterpiece of intelligent design. Its internal matrix or scaffolding (support structure) and its chromosomes are highly engineered for efficiency and dynamic reconfiguration in response to the slightest regulatory cues. At one time, scientists actually considered the nucleus to be a somewhat random, unordered environment, with the chromosomes simply assuming some sort of ambiguous orientation. We now know that this is not true at all.

The chromosomes actually occupy very specific locations and assume specific, three-dimensional structures within the cell nucleus. The organization of this assembly is aided by a complex and efficient support matrix that has attachment sites along the chromosomes and along the inside surface of the nuclear membrane. In addition to providing overall structural support to the nucleus and the chromosomes, the nuclear matrix has been shown to enhance and stabilize gene activity. Amazingly, this whole nuclear assembly is very dynamic, changing configuration based on cell type (e.g., liver, kidney, heart, etc.) and also in response to certain biochemical cues received from environmental conditions that the cell senses.

Interestingly, the three-dimensional positioning of the chromosomes within the nucleus is known to be strictly and dynamically controlled to maximize the efficien-

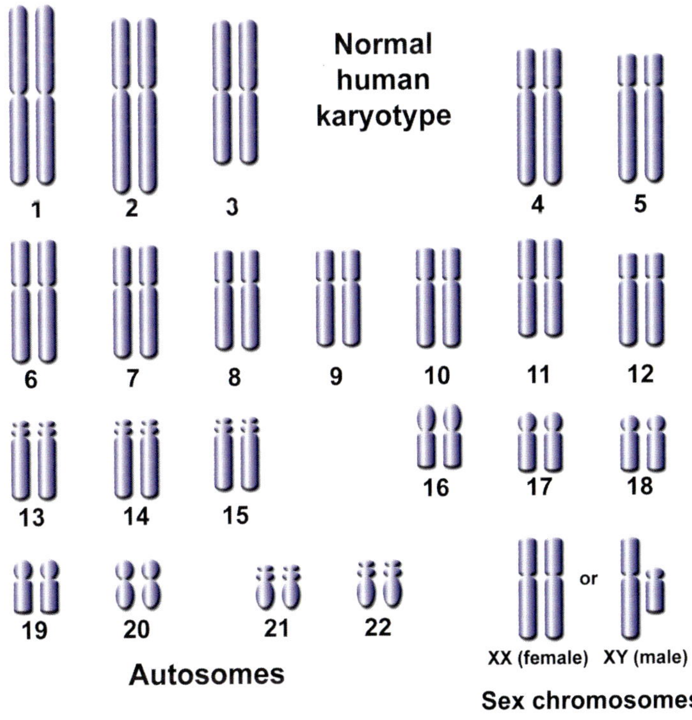

Figure 3.3 An image of what is referred to as a karyotype in which the paternal and maternal chromosome homologs are paired up and numbered appropriately. These chromosomes are of a normal human. Notice the X and Y sex chromosomes at the bottom far right; the Y is much smaller than the X.

cy of gene activity. Genes are often turned on in groups called *modules*. Therefore, the chromosomes are positioned in such a way that the genes in a module are clustered physically close to each other so that the various regulatory mechanisms associated with gene activity can be tightly interconnected for maximum efficiency. In fact, several recent studies in single and multicell eukaryotes have shown that sections of chromosomes can actually be spliced and rearranged to dynamically create important new genetic functions—all under perfect regulatory control.

As research on the function and structure of chromosomes progresses, a picture of a highly engineered instruction set that not only takes into account the turning off and on of genes, but also the specific three-dimensional structure of the genome, is beginning to emerge. This area of research is referred to as *genomics*. The nature of the genome is an excellent example of the concept of intelligent design mentioned previously. From an informational perspective, many of the same techniques and patterns employed by skilled computer programmers are evident in genetic code.

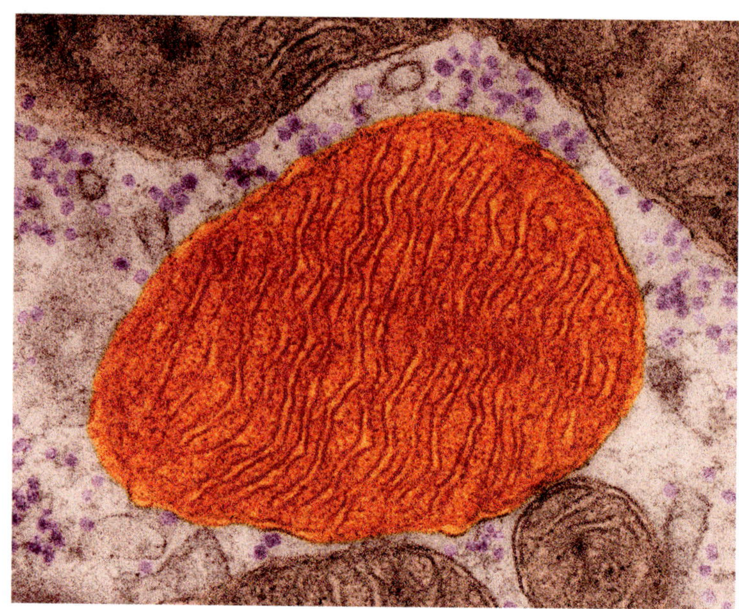

Figure 3.4 Micrograph of a mitochondrion from a heart muscle cell.

Another membrane-bound feature within eukaryotic cells is the *mitochondria*. These are highly engineered, complex organelles found in the cell cytoplasm. The mitochondria are the primary energy-producing apparatus in the cell. To maximize energy production, there are multiple numbers of these occupying the cytoplasm. Figure 3.4 contains a transmission electron micrograph image that shows that the mitochondrion is highly structured and compartmentalized, with a clearly defined inner membrane system. The primary energy molecule produced by the mitochondria and used throughout the cell is *adenosine triphosphate* (ATP). Scientists have also discovered that the mitochondria are directly involved in a regulatory manner with other important cell processes, including cell division, programmed cell death (apoptosis), signaling, and cell differentiation.

Mitochondria: The Evolutionary Model of Symbiosis Is Unsupported

Interestingly, the mitochondria have their own piece of DNA that is circular and relatively void of non-protein-coding DNA sequences, like a bacterial genome. Because the mitochondria have a circular piece of DNA that is enclosed by a membrane, scientists have used these features in an attempt to support an over-simplified evolutionary model (*endosymbiosis*) of how eukaryotic cells came to have organelles like mitochondria. The speculation involves a hypothetical scenario in which a bacterium was engulfed by a larger nucleated cell at some point in ancient time. Supposedly, the engulfed bacterium experienced a radical reduction in gene number and chromosome size, from about 6 to 7 million base pairs of DNA down to about 15,000 base pairs. This equates to a 99.7 percent reduction in important DNA-based cellular information (to 0.3 percent of the original material). That is an extremely large amount of DNA removed from the cell system, yet amidst the chaos, whatever mystical evolutionary process caused this reduction somehow kept just the right genes that the mitochondria needed.

Accompanying this marked genome depletion was the supposed loss of nearly all typical bacterial cell functions, plus the gain of new functions associated with being a mitochondrion. Other obvious difficulties with this evolutionary model are immediately evident in the overall structure of the mitochondria.

First, the compartmentalized inner membrane structure of intricate folds and convolutions (called *cristae*) and the advanced electro-biochemistry found in eukaryote mitochondria were never features found in bacteria.

Second, most of the genes that the mitochondria need to function are encoded in the nuclear DNA and not in

Figure 3.5 Hexagonal-shape plant cells from leaf tissue containing multiple chloroplasts (round green structures).

the circular piece of mitochondrial DNA. Virtually all of these nuclear genes have the hallmarks of eukaryote gene structure and do not provide support for a "gene transfer" scenario.

Third, the mitochondria do not even remotely resemble a bacterium in regard to its organelle membrane, internal matrix features, and DNA composition.

Furthermore, it definitely does not qualify as a bacterial endosymbiont (an organism living symbiotically within a host body). The mitochondria are finely tuned pieces of highly specialized machinery within a eukaryotic cell, perfectly engineered to support the internal chemistry and specific cellular functions unique to eukaryotes. It has its own piece of DNA because the specific genes encoded produce transcripts rapidly and in close proximity to the machinery where they are needed. This would not occur in an efficient way if these particular genes were located in the nucleus. The mitochondria are a finely tuned machine engineered to meet the dynamic on-demand energy production requirements in the cell.

Plant Cell Chloroplasts

The *chloroplast* is a subcellular organelle found in plant cell systems. This organelle is a key feature in the chain of life, converting light energy, carbon dioxide (CO_2), and water into carbohydrates. The carbohydrates produced by plants provide a basal food source for the earth's ecosystem, and ultimately for all types of life here that directly or indirectly eat plants. This photo-biochemical process used by plants, called *photosynthesis*, also scrubs the air of the dangerous CO_2 that results from cell respiration, a metabolic process that occurs in all life forms. When animals on earth inhale air, they take in oxygen to fuel respiration and then exhale carbon dioxide. Plants also respire and expel CO_2, mostly at night during the dark period, using the carbohydrate reserves accumulated during the day from photosynthesis.

Respiration is the process by which carbon food sources are metabolized (broken down) and converted to cellular energy with the aid of the mitochondria. Cellular energy is ultimately supplied through the energy molecule ATP, a type of cellular currency. Energy to perform a wide variety of cellular activities is released from ATP molecules when a phosphate bond is broken and ADP (adenine diphosphate) is produced.

Figure 3.5 shows a light microscope image of plant cells in which numerous green chloroplast organelles can be clearly seen. Chloroplasts occur in multiple copies within the cytoplasm of the cell to maximize photosynthesis. A typical plant cell has between 10 and 100 chloroplasts per cell. The internal structure of the chloroplast is depicted in Figure 3.6. Like the plasma membrane and the other organelle membranes, the chloroplast and its internal photosynthetic system is enclosed within a phospholipid bilayer. When scientists first examined the internal details of the chloroplast, they discovered stacks of flattened disks that were interconnected with tube-like structures. These disks are called *thylakoids*, the stacks of thylakoids are *granum*, and the interconnecting tube-like structures are *lamella*. The light-harvesting reactions that drive the photosynthetic process occur in the thylakoids.

Interestingly, chloroplasts, like mitochondria, also contain a small circular piece of DNA. The genes on the chloroplast chromosome encode some of the RNA transcripts required for photosynthesis. However, just like the mitochondria, many of the proteins required for the chloroplast to function are encoded by genes in the nuclear DNA.

The endosymbiosis evolutionary model surrounding the origin of the plant cell chloroplast is very similar to that proposed for the mitochondria. It has been proposed that at some point in ancient time, some type of precursor plant cell engulfed a photosynthetic bacterium. It is believed that the engulfed bacterium supposedly both devolved and evolved into a chloroplast in similar fashion to the hypothetical scenario described earlier for the mitochondria. The devolving would have involved a drastic loss of chromosomal DNA from about 4 to 6 million bases to about 100,000 to 150,000 bases, the typical size of plant chloroplast chromosomes. In addition, the nuclear genome would have had to gain a remarkable amount of new information to encode the additional proteins needed to create and operate a functional chloroplast. And of course, the chloroplast, in both structure and function, does not remotely resemble a bacterium—or even the remnant of a bacterium, for that matter.

Additional Eukaryotic Cell Characteristics

Another important organelle found in most eukaryotic cells is the *Golgi apparatus* (also known as the Golgi body, Golgi complex, or just Golgi; see Figure 3.7). It received its name from Italian scientist Camillo Golgi, who discovered it in 1898. The Golgi apparatus is a very important and multifunctional organelle that handles the processing of large macromolecules like proteins and lipids (oils and fats) immediately after their production.

Many molecules (e.g., proteins) need important modifications, such as the addition of small molecules like phosphate groups or metals (e.g., zinc or copper). Some even require the addition of large molecules like various types of sugars. Newly made proteins may also need to be folded with the assistance of other proteins to achieve the correct functional three-dimensional structural conformation. Translation of a messenger RNA into a linear chain of amino acids (polypeptide) or a protein is not the end of the DNA to RNA to protein process flow. There may still be a number of additional steps needed during the protein modification phase in the Golgi apparatus before the final, functional end product is achieved.

Lysosomes are specialized membrane-bound organelles found in both plant and animal cells, although they are somewhat rare in plant cells. The lysosome contains digestive enzymes that help break down a variety of materials found in the cell considered to be waste products. They do this by injecting their enzymes into the cell's vacuole, which is a specialized storage compartment in the cytoplasm. Targets for lysosome-assisted digestion could include excess or used cell organelles and their components, particles of food taken into the cell, and viruses or bacteria that have been engulfed by the cell in a process called *phagocytosis*. This type of activity is obviously potentially quite toxic to the cell, and if not compartmentalized would be impossible to perform. No evolutionary explanation for the origin and development of the lysosome exists.

Vacuoles are versatile storage structures within the cell that can interact with other cell components and organelles depending on the process undertaken. Because vacuoles have a lipid membrane structure isolating their contents from the cell cytoplasm, they are also considered to be organelles. Vacuoles can participate in a wide variety of functions, including:

1. Isolation of materials that might be harmful to the biology of the cell.
2. Maintaining cell turgor (water pressure) by osmotic gradients. This is an incredibly important feature in plants to keep them erect and healthy. A loss of cell turgor pressure in plants causes them to wilt and eventually die. Vacuoles in plant cells are often

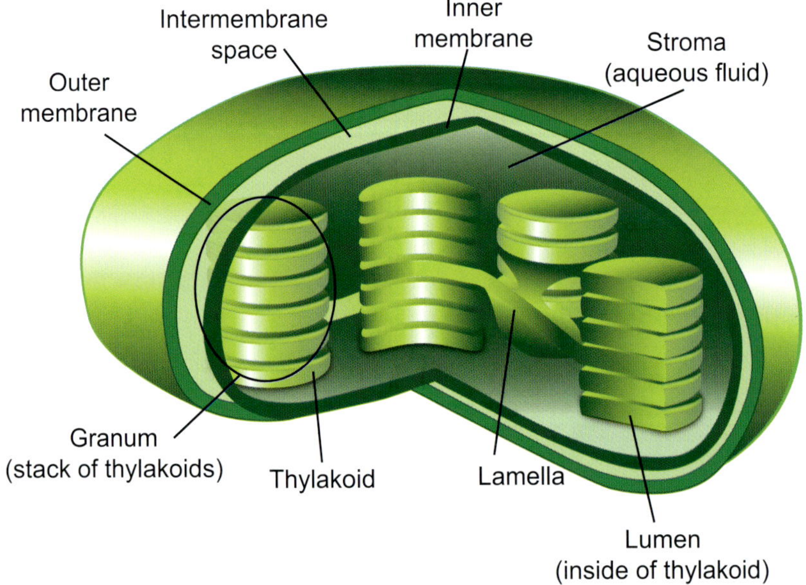

Figure 3.6 A diagram showing the internal structure of the chloroplast.

quite large and can take up a major proportion of the total cell volume (Figure 3.8). Of course, animals need to maintain a healthy level of water content in their cells, too, or they will become dehydrated, sick, and die.

3. Maintaining cell pH. Typically, more acidic levels (lower pH levels) are found in the vacuole compared to the rest of the cell. This is very important for the digestion of various food materials in collaboration with an organelle like the lysosome.
4. Housing of various molecules of importance to the cell for progressive use or storage for use at a later time. Even informational molecules like mRNAs can be stored in the vacuole for rapid translation en masse at a later time.
5. Exportation of waste products or unwanted materials from the cell. The vacuole essentially migrates to the cell's plasma membrane and expels the material through the membrane into the external environment.

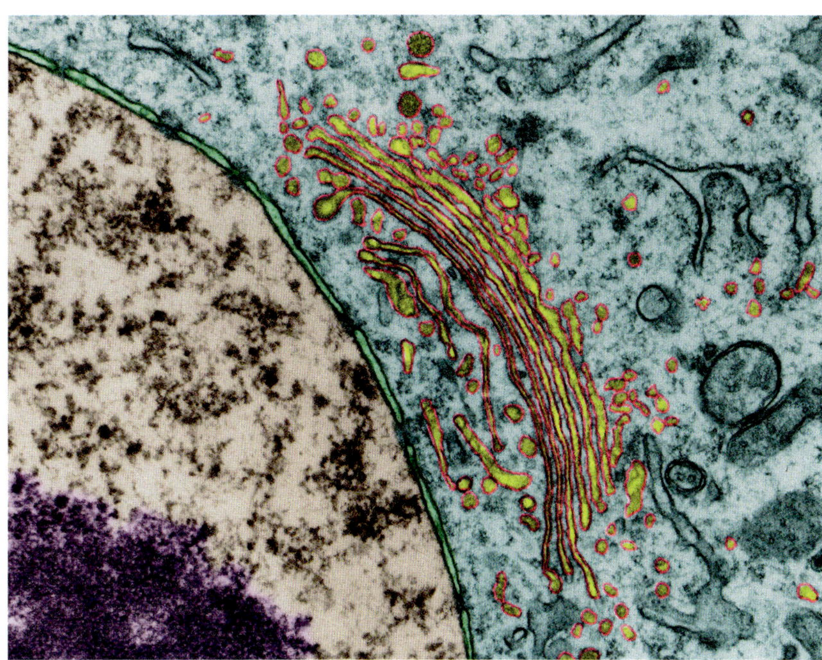

Figure 3.7 A scanning electron micrograph of a Golgi apparatus in a human leukocyte cell.

Because the vacuole is so versatile and can be involved in so many functions, its specific function depends on the organism and the location of the organ within the organism.

The *peroxisome* organelle is present in most eukaryotic cell systems. This uniquely designed intracellular apparatus contains a variety of specifically designed enzymes that are required for the oxidation of long-chain fatty acids, an important process in eukaryotic cell biology. As described previously, evolutionary reasoning typically assigns an endosymbiotic origin to some cell organelles, but it does not do so in the case of the peroxisome because of one serious conflict—there are no prokaryotes that have peroxisome-like enzymatic activity. In fact, the complicated and very important activity of the peroxisome is additional cell-based evidence for intelligent design, since it appears suddenly in higher level animal systems with no evolutionary precursor.

The chemical reactions that occur in the human and mammalian peroxisome are critical for detoxifying the body of toxic substances (e.g., peroxides) that enter the bloodstream and cells. They are also very important for the metabolism of fatty acids in the cells, a key functional component of animal life. The peroxisome is a perfect example of a highly designed cell organelle that aids in an efficient system of fault tolerance engineered into the cell system. Again, evolutionists have no clear explanation as to how the peroxisome may have come about in the eukaryotic cell. However, creation scientists see this as a perfect example of how every detail of cellular complexity is intricately engineered for optimal usefulness and purpose by the God of the Bible.

Another major subcellular structure common to virtually all eukaryotic systems is the *endoplasmic reticulum* (ER), an extensive network of membrane-based folds and sacks associated with and surrounding the nucleus. There are two general categories of ER, rough and smooth. The rough ER is studded with thousands of ribosomal protein complexes. The smooth ER is not associated with ribosomal activity, but instead hosts a wide assortment of cell processes involved in carbohydrate metabolism and in lipid and steroid production. As mentioned previously, the ribosomes are protein complexes that translate mRNAs exported from the nucleus into proteins.

In addition to a seamless association between the ER and the nucleus, it is logical that there is also a close association between the ER and the Golgi apparatus, the structure that is involved with protein modification following translation. All of the organelles involved in the process of DNA to RNA to protein production are seamlessly integrated in the cell to provide optimal efficiency of information transfer and utilization. In addition, there are multiple levels of control and regulation at all stages of the overall process.

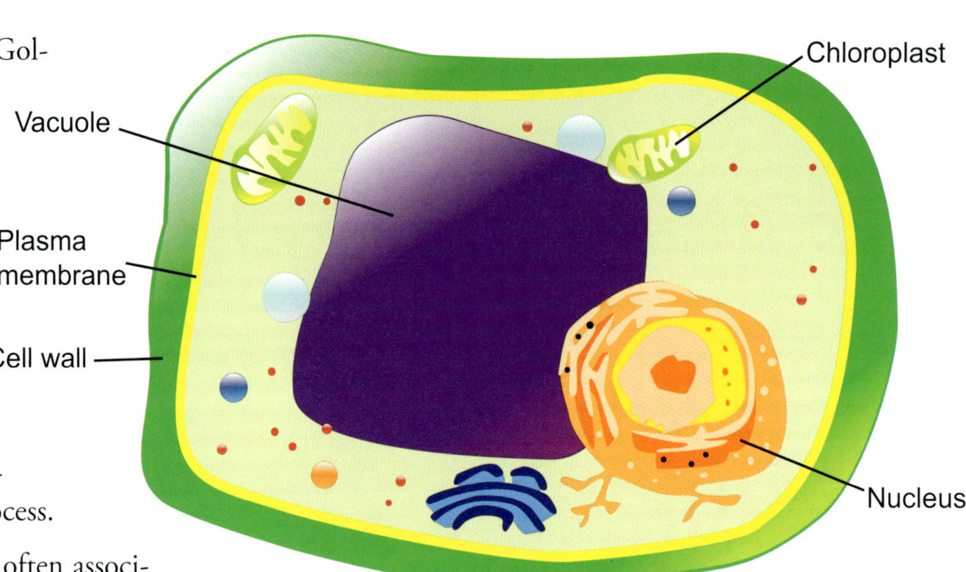

Figure 3.8 An illustration of a vacuole in a plant cell. Note the large amount of space that it occupies within the cell.

A feature that most people do not often associate with eukaryotes—although it is foundational to the cycle of life for many multicellular eukaryotes, including humans and all mammals—is the *flagellum*. The flagellum is a motorized cellular appendage used for cell motility (discussed earlier in Chapter 2). In eukaryotic cells that have flagella, the flagellum's mechanism of action and overall design pattern are completely different from that of bacteria flagella, yet it is just as efficient and useful. Once again, this is an intelligent design feature that mystifies evolutionists. Since the design of the eukaryote flagella and motor is entirely different, there is no prokaryotic evolutionary precursor for it. It appears suddenly and fully formed in eukaryotic cell systems.

In multicellular eukaryotes like mammals, the flagellum is commonly observed in the male reproductive cells (sperm), enabling them to move about in the female reproductive tract to facilitate contact with the egg cell(s). In addition to higher level eukaryotes, single-cell types such as the *Euglena*, a commonly observed pond inhabitant, also make use of flagella (Figure 3.9). Once again, a seemingly similar system is present in bacteria, but upon closer examination, it is not at all

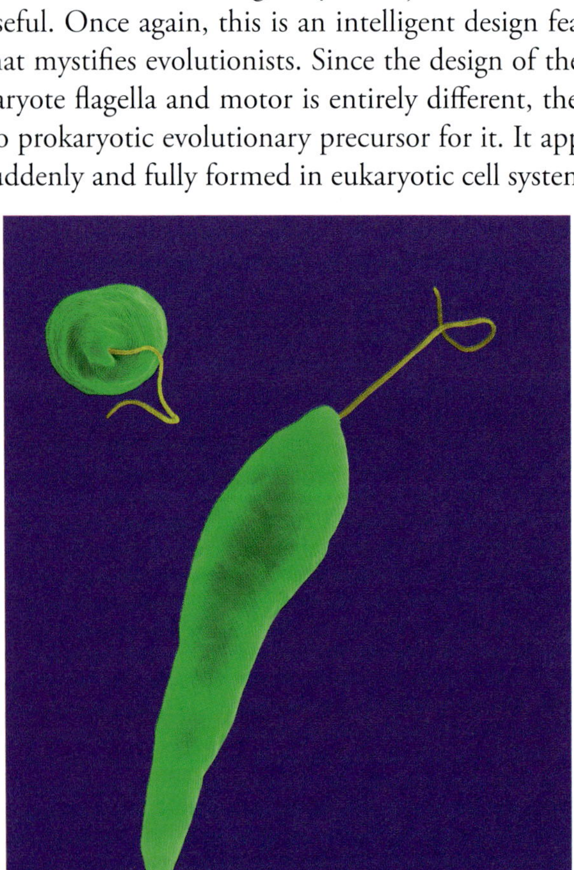

Figure 3.9 *Euglena* with flagella, a commonly observed pond single-cell eukaryote.

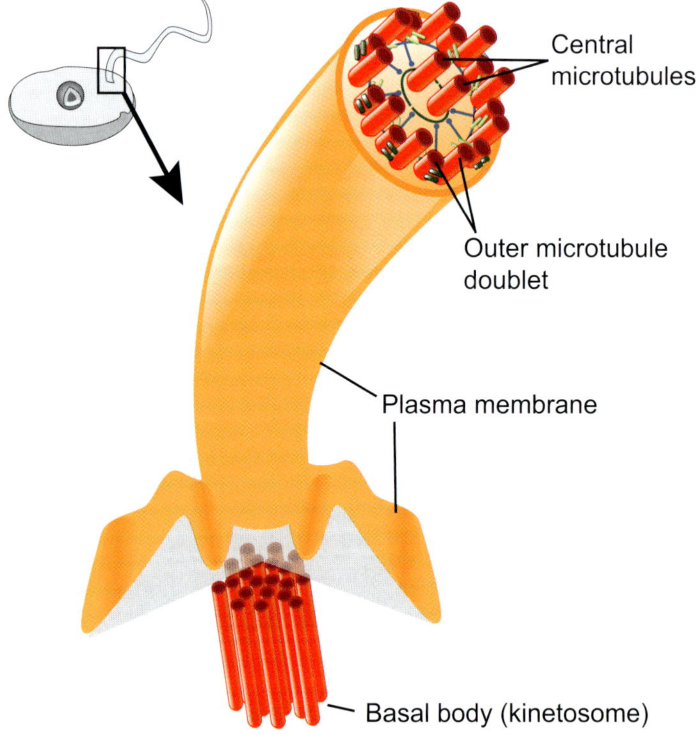

Figure 3.10 Structure of the eukaryotic flagellum.

the same and lends no credence whatsoever to the evolutionary model of a flagellum first evolving in bacteria and then becoming modified for use in eukaryotes. In fact, in the kingdom of prokaryotes, archaea have a flagellum that is completely different from either bacteria or eukaryotes.

The flagellum in a eukaryotic cell is quite a complex piece of machinery and is actually thought to do more than just propel the cell. In fact, it now appears that it serves as an apparatus for cell signaling that includes both transmission and reception of extracellular signals in the external environment. The structure is characterized by a bundle of nine fused pairs of microtubules surrounding several central single microtubules (referred to as a 9 and 2 structure; see Figure 3.10). The basal apparatus is similar to centrosomes, a structure on the inside of the cell used in other aspects of cell structure and motion, including the separation of chromosomes. The biophysical force that causes motion in the base of the structure is based on the use of ATP produced in the mitochondria, a completely different type of system than that used in bacteria. The bacterial flagellum uses a proton pump system. The flagellum structure in a eukaryote also has internal channels that allow for the transport of proteins up and down the structure that are involved in signaling, whereas in bacteria, the whip-like structure is solid.

What we normally observe at the whole organism level, we also observe at the cellular level—novel features and mechanisms in the cell appear fully formed and complete with no trace of any evolutionary precursor. Furthermore, the mechanisms and apparatus involved at the cellular level are always configured at the highest and most optimal level of engineering and efficiency—never halfway built or partially functional, as the evolutionary model would predict.

Summary

1. A *eukaryote* cell is distinguished from the *prokaryote* cell by membrane-enclosed structures that allow for compartmental functions within the cell. Although some eukaryotic organisms are single-cell structures (yeast, amoebas, etc.), most eukaryotic cells are part of multicell, highly organized, and complex creatures (mice, fruit flies, humans).
2. The *nucleus* is the command center in eukaryotic cells. It holds the informational database for the cell and sends out encoded orders as signals come in from other sources within the organism. As scientific knowledge increases about the workings within the nucleus, it is becoming more and more apparent that the nucleus is a masterpiece of design. Animal and plant cells also contain highly engineered energy-producing organelles called *mitochondria* that make complex forms of life possible.
3. *Chromosomes* are housed inside the nucleus of animal and plant cells. Chromosomes contain proteins and chains of deoxyribonucleic acid (DNA) molecules. DNA holds the instructions (the genes) that make each type of living creature unique. There are many marvelous operations within the nucleus that direct the overall function of the cell. The cell, in turn, is interrelated with similar cells, which are ultimately part of the proper functioning of the entire organism.
4. Genes are transcribed to construct proteins, many of which are taken to the *Golgi apparatus*, a multifunctional organelle that processes large macromolecules such as proteins and lipids (oils and fats). Some proteins need to be folded by other proteins into a proper three-dimensional structure. Translation of messenger RNA into a carefully constructed chain of amino acids also occurs within the Golgi apparatus. Many processes must happen just right for the cell to work. Mistakes are often disastrous!
5. Novel features and mechanisms in the cell appear fully formed and complete, with no evidence whatsoever of having gradually evolved. These mechanisms and apparatus are always configured at the highest and most optimal level of engineering and efficiency. Although many theories exist about how these marvelous designs and operations might have come about, there is absolutely no evidence of any halfway-built or partially functional stages.

Life-Giving Blood

Randy J. Guliuzza, P.E., M.D.

After 100 years, automobiles still need engine oil, transmission fluid, brake fluid, antifreeze, power-steering fluid, and so on. Wouldn't it be great if just a single multipurpose fluid could be circulated from a central reservoir? Each part would use only the needed properties of the fluid, exclude detrimental properties, and then send it back. The new system's worldwide impact would ensure a huge market—and academic honors—for the clever developers.

This lucrative breakthrough, however, would not be pioneering. Just such a brilliant integration of fluid properties to the diverse needs of the physical body has already been achieved in human blood—in a self-starting process beginning about 15 days after fertilization.

Heart and Blood Vessel Formation

The first human cell divides rapidly, becoming a small cluster that implants inside the uterus. Initially it flattens out to a disc only a few cells thick that is able to get nutrients by diffusion from maternal blood circulation. However, by two weeks after fertilization the disc is becoming too thick for this, so the developing embryo urgently needs a nutrient transport system.

Right on cue, blood and blood vessel formation begin at the end of the second week in both the embryo and developing placenta. Heart tubes (the precursor to the heart) form and start pumping within seven days. The cardiovascular system is the first organ system to become functional—an important factor, since every cell depends on blood to survive.

Vital Characteristics of Blood

Blood is a liquid tissue. For normal human function, blood has to be a fluid. Why? Fluids can flow, carry either suspended or dissolved solids and gases, and respond to even slight pressure changes by continuously changing shape. Blood and blood vessels, therefore, form an incredibly flexible conduit—the exact shape of a person's body at any moment—that connects the outside world to the deepest cells inside. Cellular metabolic demands are relentless. That is why nearly all of the estimated 60 trillion cells in the body—each one carrying out an average 10 million chemical reactions per second—are always close to blood vessels bringing oxygen and fuel.

Blood is made up of solid (formed) parts such as oxygen-carrying red blood cells (RBCs), disease-fighting white blood cells (WBCs), and platelets suspended in a liquid that is 92 percent water. This liquid, called plasma, has about 120 dissolved components that include oxygen, carbon dioxide, glucose, albumin, hormones, and antibodies. Sensors continuously monitor the concentrations of these items and make swift adjustments. Vital body functions like normal acid-base ratio, intracellular water content, blood's ability to flow through vessels, and managing body heat production depend as much on correct concentrations as the correct mix of components.

Fetal Blood Production

The embryo makes RBCs first, his most necessary blood component. These distinctive cells are made by the inner lining of blood vessels in a temporary structure outside the embryo called the yolk sac, which in people is actually a "blood forming sac" that never contains yolk. This misguided name was given because it was believed to have "arisen" in a pre-human animal ancestor and because it initially contains a yellow substance.

The progenitor RBCs eventually migrate from the yolk sac to the liver and spleen, which become the lead cell-forming sites by the mid-second month of gestation. By the fifth month, bone marrow is sufficiently formed to take over for nonstop lifelong production. Interestingly, even in adulthood if the body is stressed by a shortage of RBCs, the spleen and liver can resume production as emergency backup sites.

In children, blood formation occurs in the long bones such as the upper leg and shin. In adults, it occurs mainly in the pelvis, cranium, vertebrae, and sternum. However, development, activation, and some proliferation of certain WBCs occur in the spleen, thymus gland, and lymph nodes. Normally, sensor-control mechanisms balance mature RBCs from their production to their loss—which is about 1,200,000 cells per second. How does the marrow produce these prodigious numbers of cells?

Blood Formation: A Precisely-Planned Process

Blood formation begins with a self-renewing population of pluripotent stem cells that are capable of developing into any type of blood cell lineage (RBC, WBC, or platelet). They reproduce by making exact copies of themselves called clones or daughter cells. Some daughter cells or originals remain as pluripotent stem cells, but the rest will be "committed" to specific lineage pathways. Which cells stay as stem cells and which get committed is a random process. In contrast, the survival and expansion of cells in each lineage is precisely controlled by dozens of interacting chemical signals called colony stimulating factors (CSFs)—some produced in other body tissues. CSFs control numerous activities, including turning certain genes on and off at just the right time so each unique feature of the cells is made.

The marrow provides a protected microenvironment where immature cells grow on a meshwork of fat cells, large WBCs called macrophages, and cells lining the marrow. The meshwork compartmentalizes the nurturing process and also secretes vital CSFs. Proper growth is stimulated by strict regulation, in stepwise fashion, over both order and timing of when 12 major CSFs are introduced to the blood cells. Controls are so exact that concentrations of CSFs from other tissues can be as low as 10-12 molar (like one grain of salt dissolved in about 27,000 gallons of water). Amazingly, at certain steps in the process some of the maturing (or mature) blood cells themselves emit CSFs to direct their own development or even control the meshwork.

For RBCs, a crucial stimulating hormone is erythropoietin, commonly called EPO. Without EPO, no RBCs would be made. EPO is steadily circulated, keeping RBC production at the normal rate. But "normal" for a 10-year-old girl at sea level may not be "normal" for a 60-year-old man living on a mountain. The genes with instructions for making EPO are controlled by stimulants known as hypoxia-inducible factors (whose function depends on several vital enzymes). These factors activate EPO DNA but not in response to the number of RBCs. Rather, low oxygen concentrations induce more EPO production, which normally results in rapidly rising RBC numbers. By regulating just exactly what is needed—the blood's ability to carry adequate oxygen—the optimum number of RBCs running at maximum oxygen capacity is continuously and efficiently adjusted. Therefore, it would be fitting for EPO to be produced mainly in an organ that is very sensitive to changes in blood pressures and oxygen content, such as the renal cortex of the kidney—which it is.

Integrating Blood Properties with Organ Function

The familiar biconcave shape of human RBCs bestows the highest possible membrane surface area relative to intracellular volume and oxygen saturation rate. This makes it possible for over 250 million hemoglobin molecules in each of the billions of RBCs to be oxygen-loaded in a fraction of a second. Recall that nearly all body cells are in close proximity to blood vessels. By necessity, most of these vessels are tiny capillaries, of which 40 could be put side-by-side in the diameter of a human hair. RBCs are twice the diameter of a capillary but can actually squeeze through it. How? Structural properties in the RBC's membrane allow the cell shape to be incredibly deformed and then spring back to normal. Five specialized structural proteins confer this important ability and a genetic defect in any causes diseases due to rupturing of less-flexible RBC membranes.

RBCs are themselves living tissues. It would be possible for RBCs to consume much of their oxygen payload with little left to supply other tissues. However, RBCs have enzymes to power their metabolic processes without the use of oxygen—so they consume none of their precious cargo.

Several kinds of cells, like the clear cornea and lens of the eye, need the oxygen and nutrients carried in blood but could not function properly if coated in red blood cells. This problem is overcome by a part of the eye that acts like a blood filter. Using ultrafine portals—so small as to screen out RBCs and other proteins—a crystal-clear water-based portion carries just enough dissolved oxygen and nutrients. After nourishing the cornea, the fluid is reabsorbed—through another set of tiny holes—back into the bloodstream. Cerebral spinal fluid and urine are some other ultrafiltrates of blood in which only some of blood's properties are extracted to fill a need at a precise location.

Conclusion

From the earliest days in the mother's womb until the day of death, a person's life is in the blood. Even a person-to-person gift of blood is treasured and called "the gift of life." Human blood is indeed a gift from the Lord Jesus Christ, clearly testifying to His great creative abilities and the body's total unity of function. The Bible says that the Lord Jesus' blood is particularly special—in fact, "precious" (1 Peter 1:19)—because it is able to redeem us and cleanse us from all sin (1 John 1:9). Let us give glory "unto him that loved us, and washed us from our sins in his own blood" (Revelation 1:5).

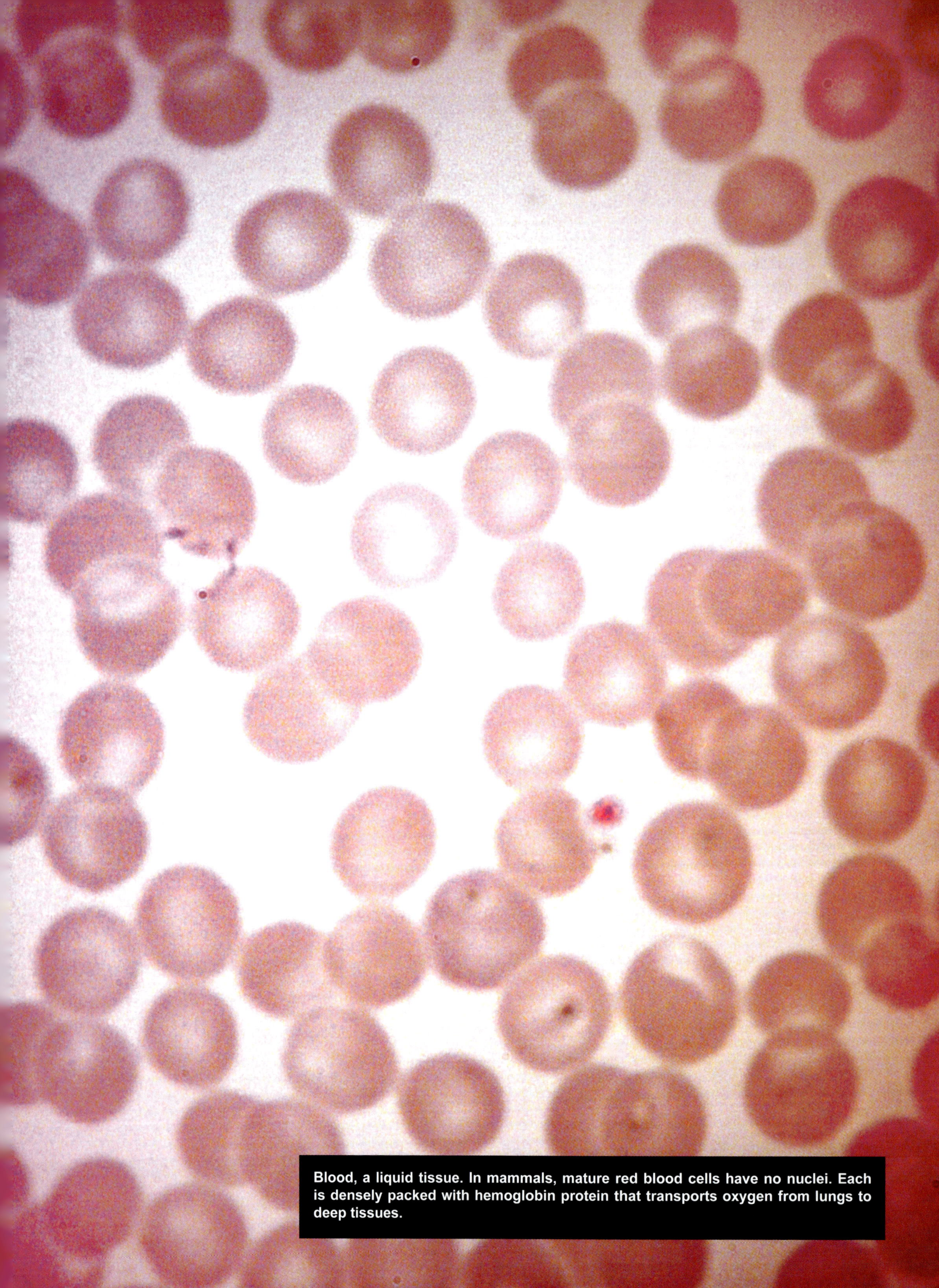

Blood, a liquid tissue. In mammals, mature red blood cells have no nuclei. Each is densely packed with hemoglobin protein that transports oxygen from lungs to deep tissues.

Chapter 4

Genes and Proteins

DNA: The Cell's Genetic Blueprint

One of the most significant components of cells is the genetic apparatus that provides the instructions and the essential informational tools needed to build a cell. All cells have chromosomes composed of deoxyribonucleic acid (DNA) that contain blocks of sequence (coded information) called genes. The genetic code is determined by the order of the base molecules adenine (A), thymine (T), guanine (G), and cytosine (C)—the molecular building blocks of DNA. These bases are connected together in a chain by a sugar (deoxyribose) phosphate backbone.

DNA naturally exists in cells as a long, double-stranded molecule. This occurs through a process called *base-pairing*. A diagram of the structure of the DNA molecule is shown in Figure 4.1. In DNA base-pairing, adenine pairs with thymine (A-T) and cytosine pairs with guanine (C-G) via a type of attraction between certain types of nucleotide molecules called *hydrogen bonding*. Although hydrogen bonds do not involve chemical changes, these molecular attractions are strong enough to keep the DNA double-stranded polymer connected and stable, like a zipper. These weak attractive bonds between complementary nucleotides, however, are not too strong to prevent specific enzymes from "unzipping" the double strands during gene activity (transcription) and DNA replication. The

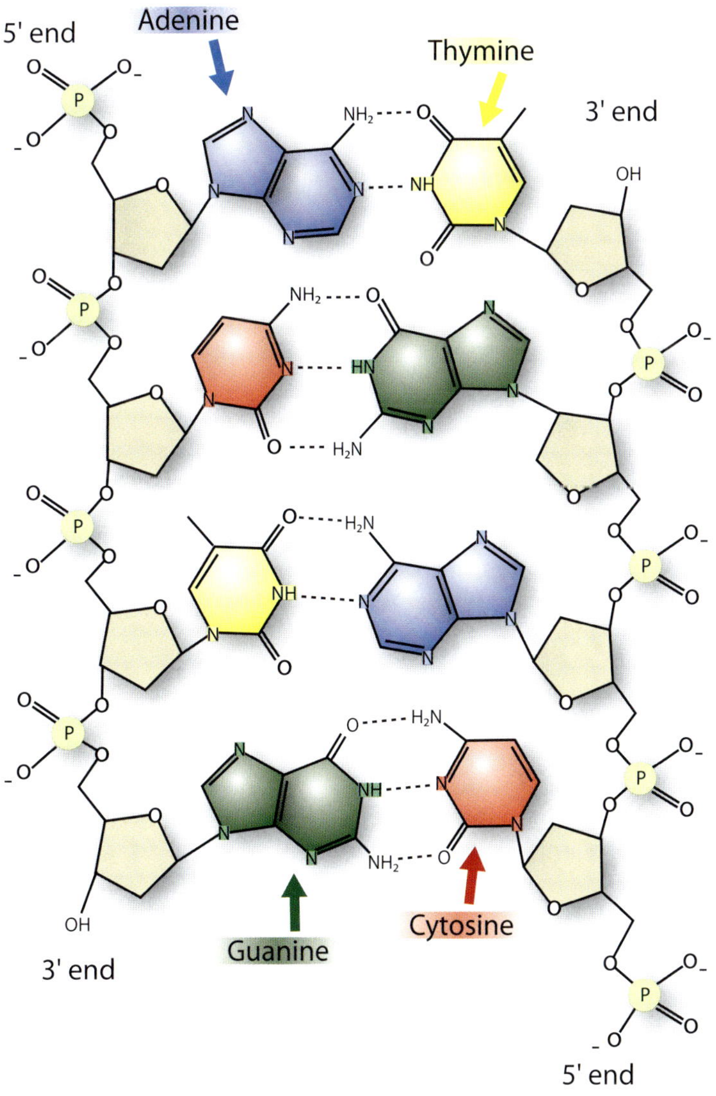

Figure 4.1 A diagram illustrating the chemical makeup and structure of DNA as it occurs in living cells.

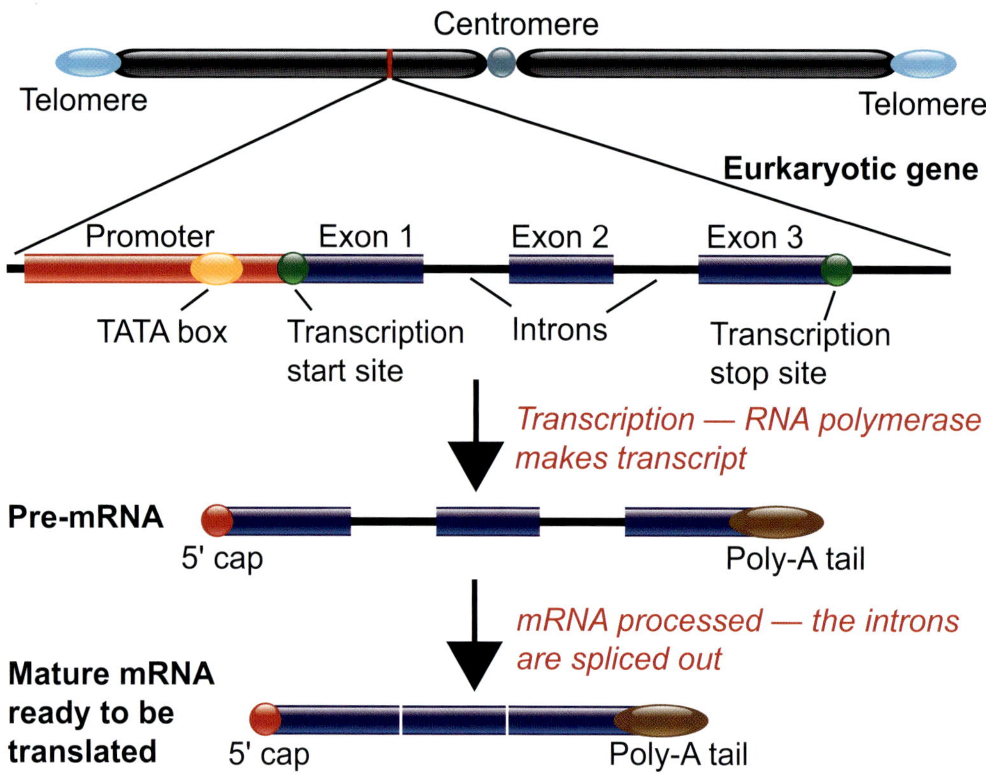

Figure 4.2 A simplified diagram showing how a gene is structured in a eukaryotic cell.

biophysical strength of the hydrogen bonds between paired bases enables DNA to properly function in many diverse processes.

A cell's chromosomal DNA is the information storage component of the cellular system and is highly integrated with the signaling and control circuitry of the cell. When production of a protein or set of proteins is required by the cell, DNA transcription is initiated by regulatory proteins that activate one or more genes. These gene-activating proteins are called *transcription factors* and bind to regulatory elements, which are short, sequence-specific DNA regions associated with genes. Many of these regulatory elements are situated in front of the gene (upstream) in a non-coding region called the *promoter* (Figure 4.2). The binding of these regulatory proteins turns the gene on or off (initiates or stops transcription), controls the rate of transcription, and also affects how the transcripts are initiated and processed. This overall cascade and orchestrated series of events is highly regulated and very complicated in many respects.

Recent advances in laboratory robotics and high-throughput technologies now allow scientists to study thousands of genes and their activity in a single experiment. They are discovering that genes are regulated in large, very complex, overlapping, and interconnected networks.

In prokaryotes, both transcription and protein production (translation) occur in the cytoplasm. Often, a ribosome will attach to a transcript while being transcribed and initiate translation. In eukaryotes, transcription occurs in the nucleus, and then the mRNA is chaperoned out of the nucleus through pores in the nuclear membrane and taken to ribosome sites in the rough endoplasmic reticulum for translation.

For the purposes of this book, we will focus on gene expression and protein production from the perspective of the eukaryotic system, using the human system as our eukaryote model (unless otherwise stated).

Making a Protein

When a gene is expressed (transcribed), a transcription factor typically binds to a special regulatory site. A key regulatory site is a DNA base sequence of TATAAA,

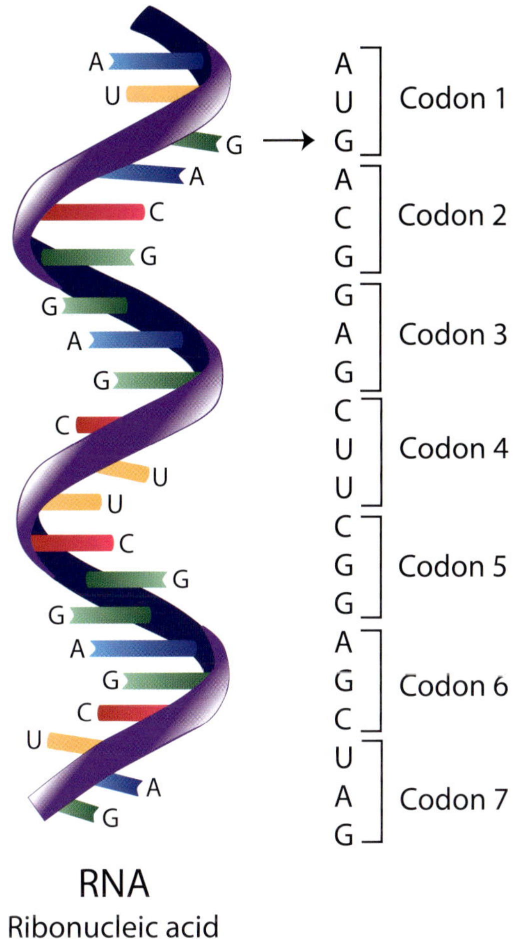

Figure 4.3 This diagram illustrates the 3-base code in an RNA transcript. Each triplet code is called a codon.

about 25 bases in front of the gene. The RNA polymerase protein complex recognizes this particular transcription factor at the TATAAA site and binds to the DNA. This initiates the process of transcription at the beginning of a DNA segment (gene) that codes for a specific protein. From this process, a copy of DNA sequence information, called an RNA transcript or a messenger RNA (mRNA), is made. The only difference between RNA and DNA is that a slightly different type of sugar is used in the connecting RNA backbone (ribose) and the base uracil (U) substitutes for thymine. The resulting mRNA is also single-stranded, whereas DNA is a double-stranded molecule.

The TATAAA sequence motif is often referred to as the TATA box. While the TATA box is probably the most ubiquitous regulatory motif in a gene promoter, in reality there are multiple types of other sequence regulation sites that occur in various combinations in a gene's promoter. Researchers have recently discovered that the combination of these diverse regulatory motifs actually helps determine how a gene is regulated.

The coding in DNA and the subsequent mRNA is dependent upon three consecutive bases called *codons* (Figure 4.3). Each codon specifies one of about 20 different amino acids. Interestingly, each amino acid can typically be specified by several different codons, and the primary informational determining feature of a codon is the first two bases. Figure 4.4 lists the various codons and their corresponding amino acids.

Once an RNA transcript is copied from a gene, it is processed in various ways. Specialized protein complexes remove and add a variety of features to an RNA transcript in a procedure called post-transcriptional processing and modification.

In prokaryotes, genes are typically represented as distinct uninterrupted units and occur in groups or clusters called *operons* (Figure 4.5). Often, all the genes in an operon will be related to a specific process, allowing them to be regulated together. For example, the well-known *lac* operon in *Escherichia coli* includes a set of genes involved in the regulation and metabolism of lactose, an important food source for *E. coli* bacteria.

In eukaryotes, genes are interrupted by non-coding segments called *introns*. The coding sections or blocks of a gene are called *exons* (see Figure 4.2). In general, introns are longer than exons. Exons average about 140 bases in humans, although one human exon was measured at over 480,000 bases. The number of exons in a gene is highly variable, ranging from 0 to over 100 depending on the gene and the organism. When an RNA transcript is made from a gene, it includes both exons and introns and is referred to as a pre-mRNA.

Immediately following transcription, the pre-mRNA is processed by the placement of protective end-caps, the removal of introns, the possible removal of some exons, and the possible addition of exons from another transcript. This mixing and matching of exons to form a mature RNA transcript that will be used to make a protein is called *alternative splicing*, referring to the alternative splicing of exons to form transcript variants derived from a single gene. As it turns out, 92 percent or more of human genes use alternative splicing to produce a variety of transcripts and, hence, multiple proteins from a single gene.

Initially, scientists were baffled by the structure of

eukaryotic genes and thought that the non-coding intron sequences served no function, leading to the term "junk DNA." However, scientists now know that introns contain many important regulatory signal sequences that control transcription and splicing events, even though introns typically do not have protein-coding DNA. In addition, some introns code for small RNAs that aide in the process of splicing and other aspects of nuclear activity.

After the RNA is processed, the mature mRNA is escorted out of the nucleus by chaperone proteins through a nuclear pore in the membrane. It is taken to the adjacent cytoplasm at a site in the rough endoplasmic reticulum that is studded with ribosomes. As cell biology research continues to lead to a better understanding of the basic aspects of single processes, design and purpose become even more readily apparent.

For example, scientists were interested in the mechanisms related to the process of moving a mature mRNA through the nucleus and into the cytoplasm. To study this, they labeled RNAs with fluorescent probes and observed their movement with a high-powered microscope. Much to their surprise, the RNAs followed a specific path as though they were on some sort of defined road or highway. RNA from a specific gene always took the same path out of the cell, although multiple paths existed depending on the gene and its location in the nucleus. The scientists had expected the RNAs to follow a random path as they were ferried through and out of the nucleus by chaperone proteins. They did not expect to witness such tight control and order, even at that basic level of cell activity.

Following the orderly removal of the mature mRNA from the nucleus, the mRNA is taken to a ribosome, engaged by the peptide-making machinery, and used as a template to make a protein. This is done by consecutively adding amino acids provided to the ribosome by transfer RNAs (tRNAs) that are charged with amino acids. This process is called *translation*. When proteins are made from mRNAs, they are said to be

nonpolar	polar	basic	acidic	(stop codon)

1st base		2nd base: U	C	A	G
U		UUU (Phe/F) Phenylalanine	UCU (Ser/S) Serine	UAU (Tyr/Y) Tyrosine	UGU (Cys/C) Cysteine
		UUC (Phe/F) Phenylalanine	UCU (Ser/S) Serine	UAC (Tyr/Y) Tyrosine	UGC (Cys/C) Cysteine
		UUA (Leu/L) Leucine	UCA (Ser/S) Serine	UAA Ochre *(Stop)*	UGA Opal *(Stop)*
		UUG (Leu/L) Leucine	UCG (Ser/S) Serine	UAG Amber *(Stop)*	UGG (Trp/W) Tryptophan
C		CUU (Leu/L) Leucine	CCU (Pro/P) Proline	CAU (His/H) Histidine	CGU (Arg/R) Arginine
		CUC (Leu/L) Leucine	CCC (Pro/P) Proline	CAC (His/H) Histidine	CGC (Arg/R) Arginine
		CUA (Leu/L) Leucine	CCA (Pro/P) Proline	CAA (Gln/Q) Glutamine	CGA (Arg/R) Arginine
		CUG (Leu/L) Leucine	CCG (Pro/P) Proline	CAG (Gln/Q) Glutamine	CGG (Arg/R) Arginine
A		AUU (Ile/I) Isoleucine	ACU (Thr/T) Threonine	AAU (Asn/N) Asparagine	AGU (Ser/S) Serine
		AUC (Ile/I) Isoleucine	ACC (Thr/T) Threonine	AAC (Asn/N) Asparagine	AGC (Ser/S) Serine
		AUA (Ile/I) Isoleucine	ACA (Thr/T) Threonine	AAA (Lys/K) Lysine	AGA (Arg/R) Arginine
		AUG[A] (Met/M) Methionine	ACG (Thr/T) Threonine	AAG (Lys/K) Lysine	AGG (Arg/R) Arginine
G		GUU (Val/V) Valine	GCU (Ala/A) Alanine	GAU (Asp/D) Aspartic acid	GGU (Gly/G) Glycine
		GUC (Val/V) Valine	GCC (Ala/A) Alanine	GAC (Asp/D) Aspartic acid	GGC (Gly/G) Glycine
		GUA (Val/V) Valine	GCA (Ala/A) Alanine	GAA (Glu/E) Glutamic acid	GGA (Gly/G) Glycine
		GUG (Val/V) Valine	GCG (Ala/A) Alanine	GAG (Glu/E) Glutamic acid	GGG (Gly/G) Glycine

Figure 4.4 The codons and their corresponding amino acids.

A. Structure of a bacterial operon

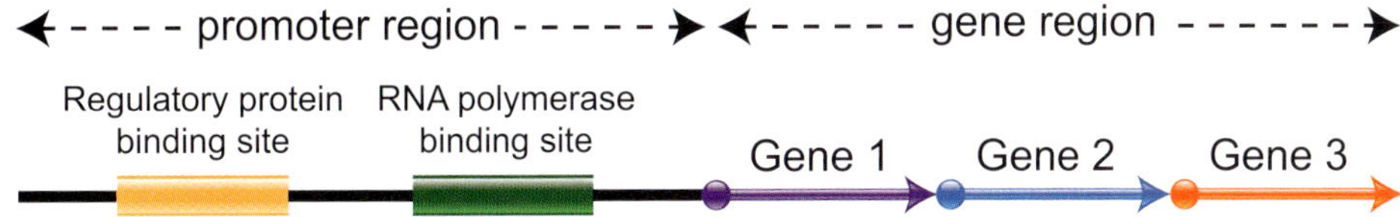

B. Expression of a bacterial operon

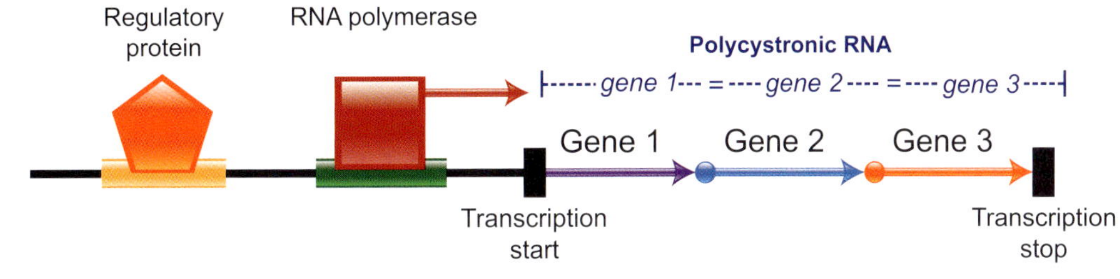

Figure 4.5 A diagram depicting the structure of a polycystronic gene in a bacterial genome.

translated. Ribosomes are large, complicated protein complexes that perform amino acid polymerization by consecutively transferring the correct codon-specified amino acid from a complementary codon presented by a charged tRNA.

The final protein product (mature protein) is not merely a chain of amino acids, but a unique three-dimensional structure folded and shaped in specific ways. Most proteins are further processed after translation by protein-assisted folding. In many cases, they are also modified chemically via the addition of metal ions or sugar molecules, are complexed with other proteins, or are perhaps associated with lipids in membranes. Most of the protein modification and post-translational changes take place in the Golgi apparatus to produce a mature, fully functional protein.

Once fully processed, this protein is transported to its destination and functional site in the cell. Much of the transportation in the cell is done by devices called *vesicles*. Vesicles are lipid bilayer enclosures (membranes) that are flexible, portable, and easily able to merge with many other similarly bound cellular compartments, enabling the efficient transfer of all sorts of cellular cargo. For the most part, proteins perform the majority of work in the cell and also provide many structural features. There are many different types and classes of proteins, and the diversity of protein biochemistry is beyond imagining.

A Masterpiece of Design and Purpose

When the human genome was first sequenced to a level where scientists began to realize how many genes were present, they were completely shocked. To their amazement, it became evident that the human genome had no more than 30,000 genes total. Eventually that number was even more refined as genome sequencing continued to progress. It is now known that there are around 23,000 genes in the human genome. This is significant for a couple of reasons.

First, this number of genes is comparable to other animal genomes, even though humans are at the top of the supposed evolutionary tree of life and we consider ourselves significantly more biologically complex than other organisms.

Secondly, over the years, scientists have discovered and characterized over two million protein variants in the human system. Therefore, there is a massive disparity between the number of genes identified and the number of proteins discovered. As the phenomenon of alternative splicing and complex gene regulation became better understood, these data began to make sense, especially within a creation-based framework. The genome is so efficiently coded that only 23,000 genes are able to produce two million different proteins. Of course, the first question that must be answered in the secular research community in light of all these findings in the field of genomics is: "How did this level of informational efficiency and immense networked complexity arise by chance random processes?"

Evolutionary bias originally prompted scientists to label introns as junk DNA because their system of thought did not allow for viewing features of the cell as having design and purpose. In some ways, one must wonder if this type of philosophical approach to science may actually be a hindrance to new discoveries in cell biology. On the other hand, a biblical creation worldview would prompt a scientist to inquire about some unknown aspect of the cell within the framework of "What did God design this to do—what is its purpose?" How much more would we know about genomics and genetics if this were the prevalent view in our scientific research community?

Summary

1. DNA contains blocks of coded information called *genes*. The *code* lies in the linear arrangement (sequence) of the base molecules composed of adenine (A), thymine (T), guanine (G), and cytosine (C). A sugar phosphate backbone connects the base molecules in a long, double-stranded chain, with the base molecules always pairing as A-T and C-G. The required precision of the DNA strand is obvious.
2. Genes are *regulated* in large, very complex, overlapping, and interconnected networks. Gene activity is called *transcription*. Working within specific sections of the DNA, specialized proteins called *transcription factors* bind with *regulatory proteins* that control the rate of transcription by turning the gene on and off. The engineering of this critical process is executed with minute timing.
3. *Messenger RNA* (mRNA) is responsible for transferring information from the genes in the DNA with a "message" to be developed into a very specific protein needed in the function of the cell. The mRNA, developed in the nucleus under carefully controlled processes, ultimately brings its message to the *ribosomes*, where the needed proteins are constructed. Every single step in this process must be tightly regulated or the cell dies.
4. Messenger RNAs from specific genes always take specific pathways out of the nucleus. Scientists were very surprised to see such tight control and order at this basic level. Evolutionary thought would have expected random operation. Precision of this magnitude would require design.
5. Human DNA contains around 23,000 genes that make up the human *genome*. Those genes are responsible for producing over two million different proteins. This enormous level of *information* is precisely *transferred* and *efficiently delivered*. The answers to the many questions about the development of these complex processes are shaped by the presuppositional assumptions of the scientist. Evolutionary naturalism must attempt to shape an answer that excludes supernatural power and intelligence. Biblical creationists would insist that the obviously engineered design and precise efficiency of the processes demand an omniscient and omnipotent Creator.

Where Did Flesh-Eating Bacteria Come From?

Brian Thomas, M.S.

Flesh-eating zombies may be the work of science fiction horror, but necrotizing soft tissue infection—a severe type of infection that destroys tissue—is a real condition that can kill about 30 percent of those infected and disfigure the rest.

Such tissue infections are rare and involve bacteria growing inside the body and often deep under the skin. They are difficult to diagnose because many cases show little outward signs. These infections progress rapidly, so quick intervention is important in increasing the patient's chances for survival. Researchers at the Los Angeles Biomedical Research Institute have found that simple blood tests, such as white blood cell count and serum sodium level, helped physicians to determine whether patients have the disease and to decide on a treatment protocol.[1]

How can flesh-eating bacteria exist in a creation once deemed good by its Creator?[2] Scientific observation is consistent with the concept that many infections, including those resulting from flesh-eating bacteria, are products of healthy cell parts that have broken down. Several different bacteria—including *Streptococcus pyogenes, Vibrio vulnificus,* and MRSA[3]—can be responsible, singly or together, for causing necrotizing soft tissue infections. These bacteria are able to counteract the body's defense systems with various molecules called "virulence factors."

In some cases, the virulence factors could have originally been good molecules that were used by bacteria to compete for survival space, thus maintaining a healthy ecological balance. Others could have associated with higher organisms in neutral, or even beneficial, ways. The *Vibrio* bacteria, some of which are responsible for cholera, produce molecules that interact specifically with epithelium in fish and squid pouches in a delicate coordination to produce bioluminescence for hunting. The proteins from *Vibrio* are only virulence factors when they are found in inappropriate environments like the human body.[4]

Some molecular machinery that enables infections today could also have originally been made for specified purposes, but may have since degenerated to the point that it can interact with other organisms, thus causing infection. Anthrax would fall into this category.[5] In addition, the Creator may have provided these "toxins" in part for their medicinal potential, as in the case of botulin, knowing that they would be needed in a fallen world.[6] In fact, there is now a patent on the use of *Streptococcus pyogenes'* main virulent factor to treat connective tissue disorders.[7]

While it is not yet known what best explains the presence of flesh-eating bacteria, current scientific observation is consistent with the fall of creation, as recorded in Genesis.

References

1. Blood tests can help detect presence of necrotizing soft tissue infections. Los Angeles Biomedical Research Institute press release, December 4, 2008.
2. Genesis 1:31.
3. Thomas, B. New Antibiotic Kills Drug-resistant Superbugs. *ICR News.* Posted on icr.org July 14, 2008, accessed December 5, 2008.
4. Sherwin, F. 2005.Creation, Corruption, and Cholera. *Acts & Facts* 34.(11).
5. Wood, T. 2002. The Terror of Anthrax in a Degrading Creation. *Acts & Facts* 31 (3).
6. Romans 8:22-23.
7. Method of treatment of connective tissue disorders by administration of streptolysin O. U.S. Patent 6998121. Posted on freepatentsonline.com.

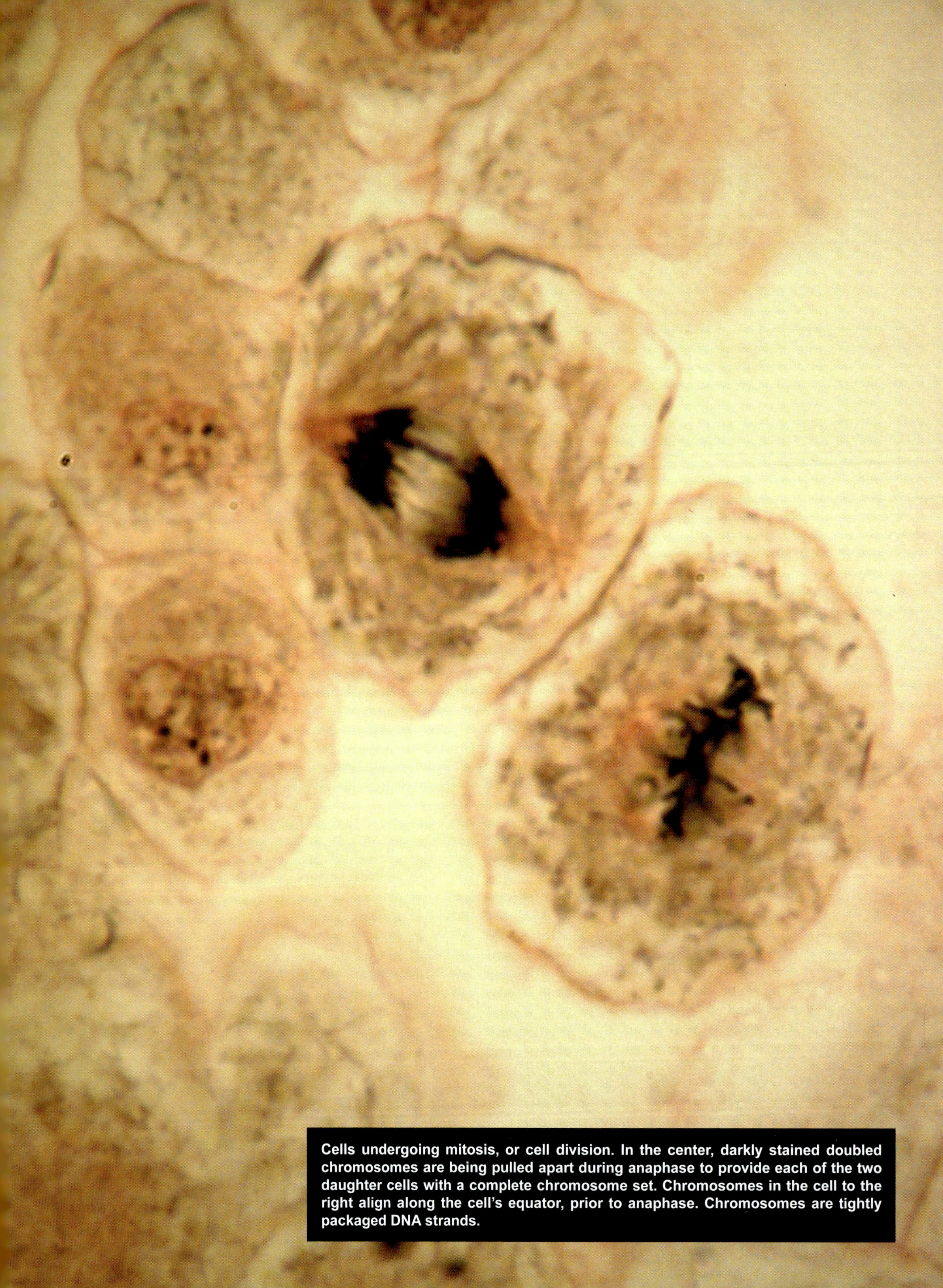

Cells undergoing mitosis, or cell division. In the center, darkly stained doubled chromosomes are being pulled apart during anaphase to provide each of the two daughter cells with a complete chromosome set. Chromosomes in the cell to the right align along the cell's equator, prior to anaphase. Chromosomes are tightly packaged DNA strands.

Chapter 5

Cell Division and DNA Replication: How Life Is Engineered to Perpetuate

In eukaryotic systems, cells can generally be classified as either being in a state of interphase (static stage) or in a state of mitosis (actively dividing). The interphase is common for fully differentiated cells that have a function to perform in a specific organ. In other words, these cells have achieved their final mature forms and are now performing functions within an organ or tissue. In some cases, this state may only last for a short period of time before these cells die and are replaced. In other cases, such as neuron cells in the brain, they may live a long time after differentiating. Most errors cannot be tolerated in cell division because of the severe complications that can develop. DNA replication prior to cell division is a process that requires extreme accuracy and fidelity for the information to be passed along uncorrupted. When random base changes occur, especially in genes that control the cell cycle, diseases like cancer can develop in which cells proliferate out of control.

Cell Division: The Basic Biological Process Required for Life

Cell division is an incredibly important process in the development, growth, and maintenance of an organism. It is the biological basis for how organisms develop, grow, function, and perpetuate themselves. The process of cell division is another impossible hurdle for the evolutionary model, not only because of its incredible complexity, but also because of its immeasurably precise nature.

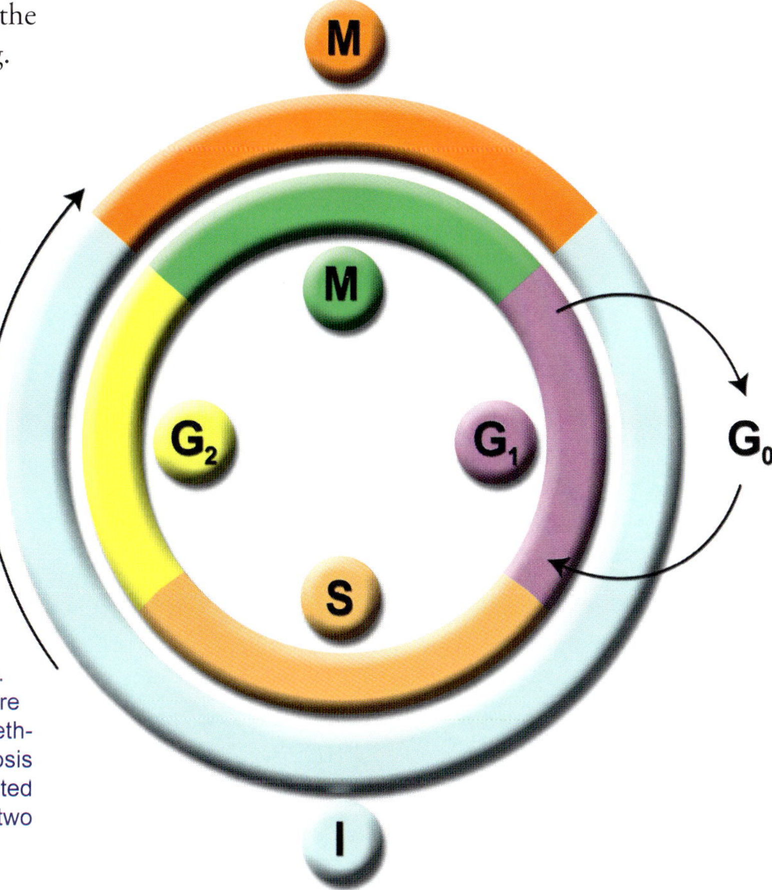

Figure 5.1 The Cell Cycle Stages: When the cell is in a state of interphase (I), before undergoing mitosis it increases in size (G_0) and exhibits a checkpoint (G_1) to determine if the cell is ready to enter the DNA replication phase. During the DNA synthesis phase (S), the chromosomes are replicated. Another checkpoint phase (G_2) determines whether the cell is ready to enter mitosis and divide. In the mitosis phase (M), the newly replicated chromosomes are separated into two new nuclei and the cell divides (cytokinesis) into two new daughter cells.

In between DNA replication and cell division (cytokinesis), a series of elaborate events occurs through which the chromosomal material (old and new) is sorted and distributed into two new nuclei. This process is called *mitosis*. Figure 5.1 illustrates the stages of DNA replication, mitosis, and cytokinesis that result in the generation of two daughter cells.

Consider that a human cell has two sets of chromosomes, each with 3 billion base pairs of DNA, for a total of 6 billion bases. This massive amount of information must be replicated quickly and accurately every time a cell needs to divide. Amazingly, this process occurs continuously in the body. When the DNA is replicated, the two strands are unzipped at multiple locations up and down the chromosome. Special enzyme complexes called *DNA polymerases* are then recruited to these open strands, where they lock down on the DNA and replicate a new strand using the old strand as a template. These open pockets in the double-stranded DNA where the polymerase complex is moving (in both directions) are called *replication forks*. The result of this replication process is two sets of chromosomes made up of both old (template) DNA and newly replicated DNA. Because each new double-stranded DNA molecule is made of one old strand and one new strand, it is called *semi-conservative replication* (Figure 5.2). The process is somewhat similar in bacteria, but considerably less complicated and extensive because bacteria typically only have one small (comparatively) circular chromosome.

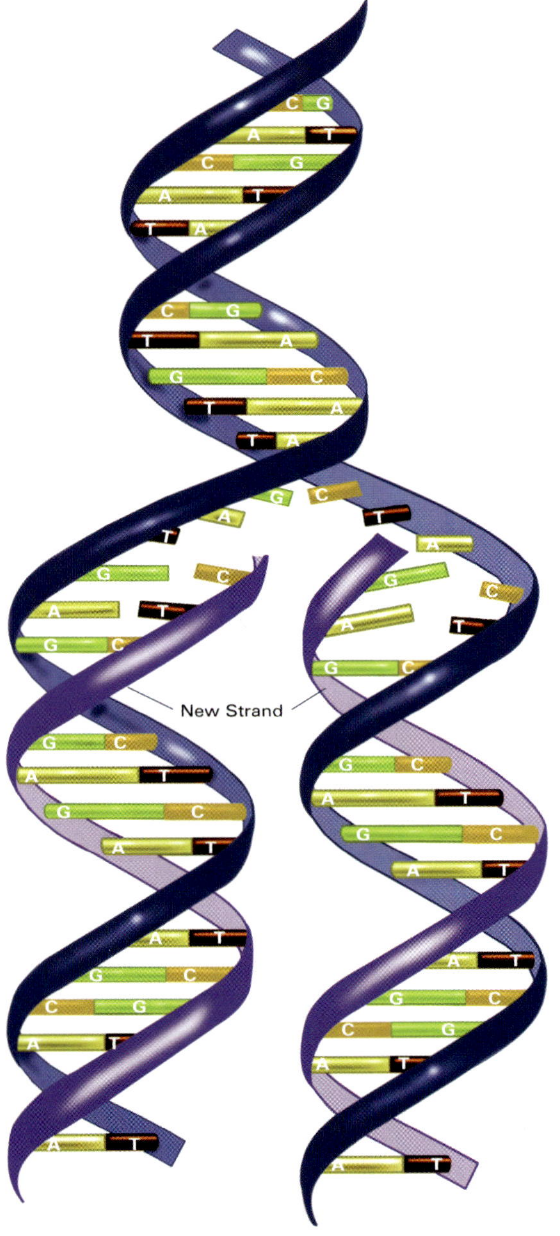

Figure 5.2 An image depicting the concept of semi-conservative DNA replication, where each new double-stranded DNA molecule is composed of one old strand and one new strand.

Cell division typically falls into two general categories. The first category is called *somatic cell division* and involves all the tissues and organs in the body that are not associated with making sperm or egg cells. Some tissues have considerably more cell division occurring than others, such as the bone marrow that continually generates new cells. Other tissues, such as brain tissue, tend be quite static and have very low levels of cell division. Of course, when an embryo, infant, or child is growing, the levels of cell division in these developing bodies are much higher than in an adult individual.

As was stated previously, one of the most critical and complex features of cell division involves the replication and separation of DNA, or the cell's chromosomes. It is not just sufficient that the DNA be replicated accurately, but the extra chromosomes have to be properly sorted, divided, and partitioned within the cell. This involves many complex activities that control how chromosomes are packaged, are positioned in the nucleus and cell, move around, and are formed into two daughter nuclei. Scientists have called this process mitosis (Figure 5.3) and have documented specific stages during mitosis based on the incredible events that take place. Below is a brief summary of the various stages as mitosis progresses.

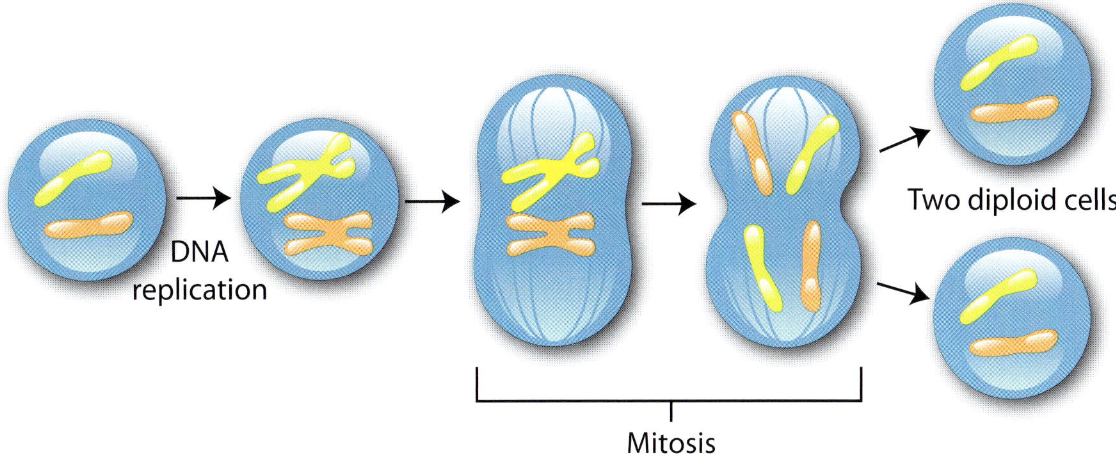

Figure 5.3 The process of mitosis whereby newly replicated chromosomes are separated into two new nuclei and two new daughter cells are then formed.

Mitosis

Scientists consider there to be several general stages of mitosis that describe how the process works. While virtually all cell processes are examples of interdependent systems and irreducible complexity, the process of mitosis and meiosis (to be described next) are perfect and dramatic examples that emphasize this concept.

Interphase: This is not a stage involved in cell division. Instead, interphase is the term used for a cell that is not dividing (undergoing mitosis), but that is rather performing its regular intended function.

Prophase: Immediately prior to prophase, the cell replicates its DNA in preparation for mitosis. There are now two sets of chromosomes that must be partitioned into two new daughter cells. This is the central, but rather complicated, goal of mitosis. Considering that animals and plants can have, in some cases, over 100 chromosomes, this is no simple task. During prophase, the chromosomes condense and actually become visible with staining under a light microscope. Many of the pictures typically shown of chromosomes are at this stage. However, this phase is very brief and only occurs in dividing cells.

Chromosomes in non-dividing cells are actually considerably more spread out, less packaged, and very difficult to see. The tight packaging that the chromosomes undergo during cell division facilitates all of the activities that occur during mitosis. During prophase, the nuclear membrane is also dissolved.

Metaphase: The highly condensed chromosomes line up in the center of the cell along a linear axis called the metaphase plate or equatorial plane. They are positioned by the spindle fibers called *centromeres* (a type of microtubule) attached to places on the chromosome, which can appear as restricted or pinched areas under a microscope during chromosomal condensation in mitosis. A protein structure called a *kinetechore* binds to the centromere as an attachment point for the spindle fibers.

Anaphase: The spindle fibers pull the two sets of chromosomes apart from each other to opposite poles in the cell. At this point, the cell starts to elongate in preparation for cell division.

While scientists have begun to characterize many of the processes that occur in mitosis, one of the greatest mysteries is how the cell knows the difference between chromosomes so that it can properly separate them and pull the appropriate ones to the correct locations in the dividing cell. There is obviously an incredible amount of information that must be communicated to the cell machinery that is involved in this whole process. Where this information comes from and how it is utilized are the big questions.

We are just beginning to understand how the hard-coded information in DNA is utilized in the cell, but it also appears that information is somehow transferred and stored temporarily in the system configuration programming of the cell matrix. We know this to be true because cells that have been programmed still follow the same programmed developmental or physiological cascade even after they have had their nucleus removed

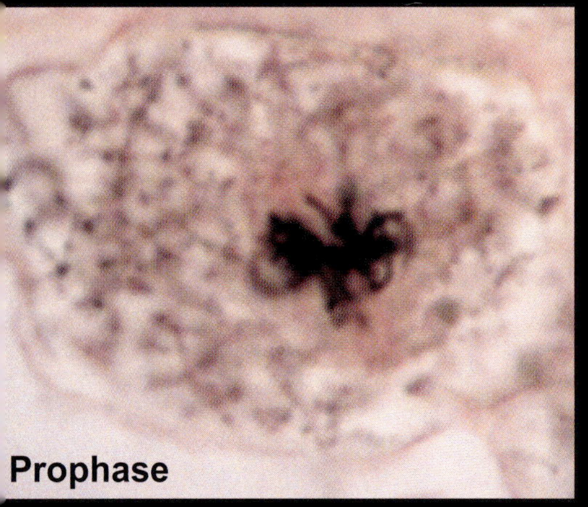

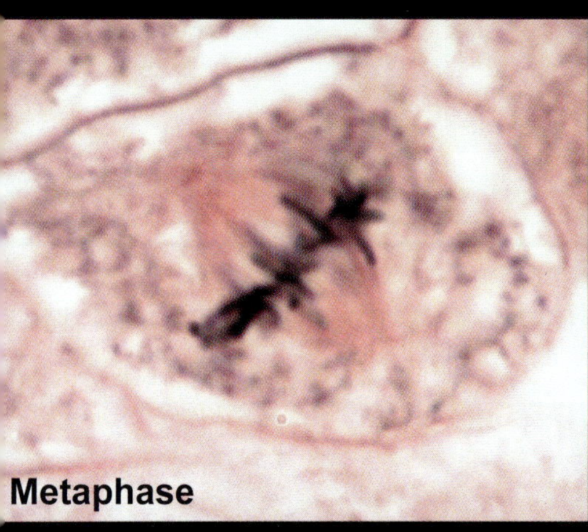

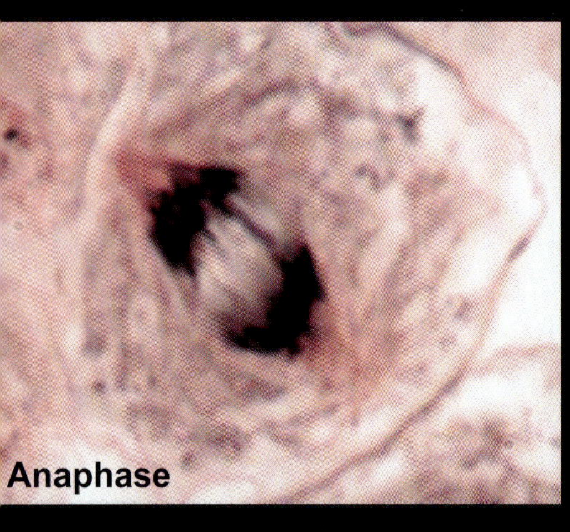

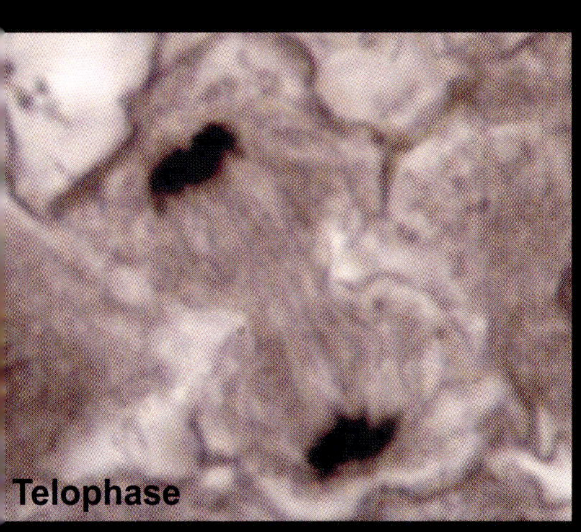

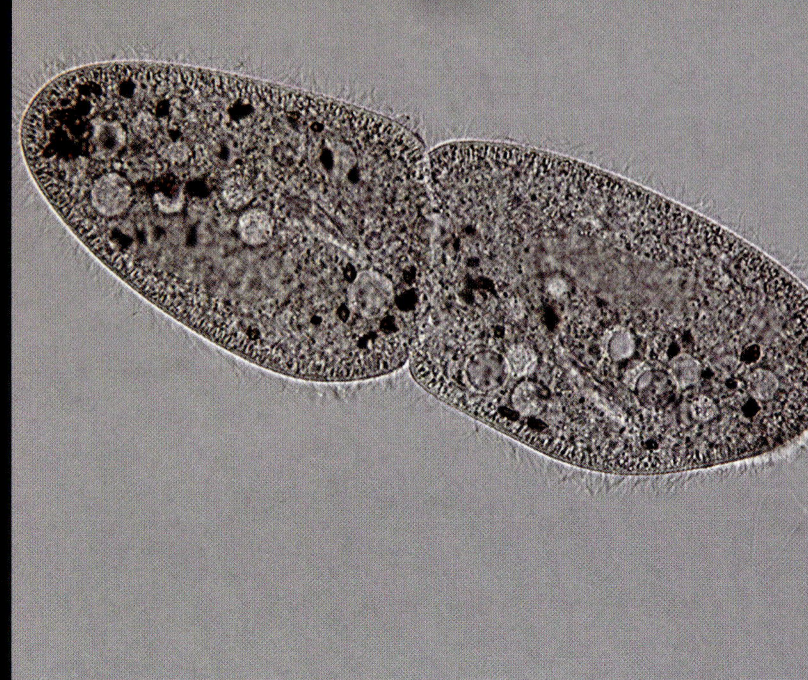

Figure 5.4 A microscopic image of two cells in the final stage of cytokinesis.

and a foreign nucleus implanted.

Telophase: The chromosomes arrive at the poles and the cell splits into two daughter cells, each with a new nuclear membrane that has formed around the chromosomes. Once again, this is an absolutely amazing and little-understood process. Not only are the chromosomes partitioned properly to each side of the cell, but they are then formed into a very specific three-dimensional structure surrounded by a new nuclear membrane and endoplasmic reticulum.

Cytokinesis: This is not an official stage of mitosis, but the term relates to mitosis and cell division. It refers to the actual splitting of the cell and its cytoplasm at the end of mitosis. It begins with a developing cleavage furrow (evident in microscopic images) that progresses to the actual splitting into two daughter cells complete with intact and functional plasma membranes (Figure 5.4).

Meiosis

The other type of cell division that occurs in animal and plant systems is called *meiosis*. Meiosis occurs during the formation of cells known scientifically as gametes, specifically sperm and egg cells. The overall process of gamete production is actually called *gametogenesis*, with meiosis being a critical process or component of it.

Meiosis has essentially two missions in facilitating gametogenesis. The first is genome reduction whereby the chromosome content—diploid (2N) in most cases, with two complete sets of chromosomes—is reduced to a haploid (1N) with one set. When a haploid egg and a haploid sperm combine, the diploid or 2N normal state of the genome is reconstituted. This is why people have two sets of chromosomes, one from the father (paternal genome) and one from the mother (maternal genome).

The second goal of the meiotic process is to create genetic variation, like shuffling a deck of cards. Maintaining genetic diversity or variation is a key feature in maintaining animal and human health. The opposite of this concept is demonstrated by the higher rate of birth defects commonly observed where genetic diversity is lost due to inbreeding.

The process of creating genetic variation in meiosis is actually quite ingenious and intriguing, since it involves two separate types of randomization/shuffling. The first phase of randomization is a form of literal genome shuffling on a massive scale. In fact, it is an absolutely amazing event that could be completely fraught with extreme hazard if governed by anything short of the most precise engineering. This event is referred to as *recombination*, and first involves the pairing up of homologous chromosomes. In other words, chromosome number 1 inherited from the father literally pairs up with chromosome number 1 inherited from the mother, and so on.

In the human genome, there are 22 regular chromosomes (autosomes) that pair up in a perfect fashion. Once paired, the homologous chromosomes literally begin "crossing over" each other. Highly efficient cell machinery begins the process of slicing, dicing, and reconnecting fragments of DNA back and forth between the paired chromosomes in what is thought to be a random but controlled manner. The end result is that the maternal and paternal chromosomes become new chimeric or recombined chromosomes, having exchanged numerous segments with each other. Although the genes and various other DNA features have been exchanged/recombined, the linear order of these components is maintained throughout the whole process. This is one of the remarkable but necessary attributes of recombination that maintain genome stability and function.

It should be noted at this point that the sex chromosomes in mammals, and humans in particular, are completely different from each other (X and Y chromosomes, see Figure 5.5). Although they are considered homologs, they undergo very little to no recombination. In humans, males have an XY makeup, while females have an XX complement. Since only one X chromosome can be activated in a normal human cell, females will have one X chromosome that is largely inactivated.

The second level of genetic variation is created in the phase immediately following recombination, when the chromosomes are pulled apart to form two separate sets. This occurs in a process called independent assortment, meaning that a newly recombined chromosome has a 50 percent chance of ending up in one set or the other. This offers a second level of randomization, helping to increase and maximize the level of available genetic diversity that can be created by the process. This

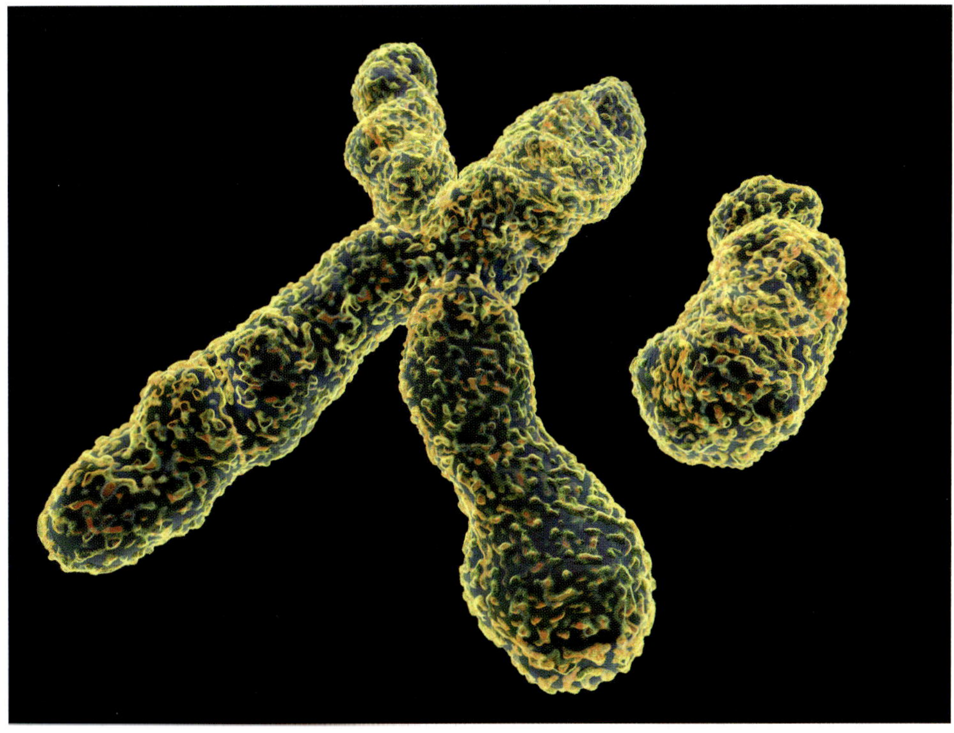

Figure 5.5 An image of the human X and Y chromosomes illustrating the major difference in size.

randomization process, in which the genetic deck of cards is shuffled twice, is why in most cases siblings look different even though they have the same parents.

Meiosis, in contrast to mitosis, is basically a one-way continuous process, and is not part of a stop-and-go cell cycle system like mitosis. In addition, meiosis only takes place in the reproductive organs of plants and animals, not in the other tissues. Its whole goal is to create new genome combinations and promote the genetic diversity needed to safeguard against the effects of inbreeding. Of course, meiosis is also the biological reason why everyone is unique.

One can also consider the process of meiosis to be the important factor enabling the fulfillment of God's dominion mandate. After both creation and the Genesis Flood, God issued commands for His living creatures to be fruitful, multiply, and to fill or replenish the earth.

The dominion mandate issued after creation:

> And God blessed them, saying, Be fruitful, and multiply, and fill the waters in the seas, and let fowl multiply in the earth. (Genesis 1:22)
>
> And God blessed them, and God said unto them, Be fruitful, and multiply, and replenish the earth, and subdue it: and have dominion over the fish of the sea, and over the fowl of the air, and over every living thing that moveth upon the earth. (Genesis 1:28)

The dominion mandate issued after the Flood:

> Bring forth with thee every living thing that is with thee, of all flesh, both of fowl, and of cattle, and of every creeping thing that creepeth upon the earth; that they may breed abundantly in the earth, and be fruitful, and multiply upon the earth. (Genesis 8:17)
>
> And God blessed Noah and his sons, and said unto them, Be fruitful, and multiply, and replenish the earth. (Genesis 9:1)
>
> And you, be ye fruitful, and multiply; bring forth abundantly in the earth, and multiply therein. (Genesis 9:7)

It is now well-established that genetic diversity and the supporting biological mechanisms are key factors allowing plants and animals to spread out geographically, fill new niches, and establish adapted, stable populations. The many irreducibly complex features and aspects of both mitosis and meiosis provide resoundingly strong arguments for special creation.

Unlike mitosis, meiosis undergoes two sets of cell division instead of one, and is separated into two parts:

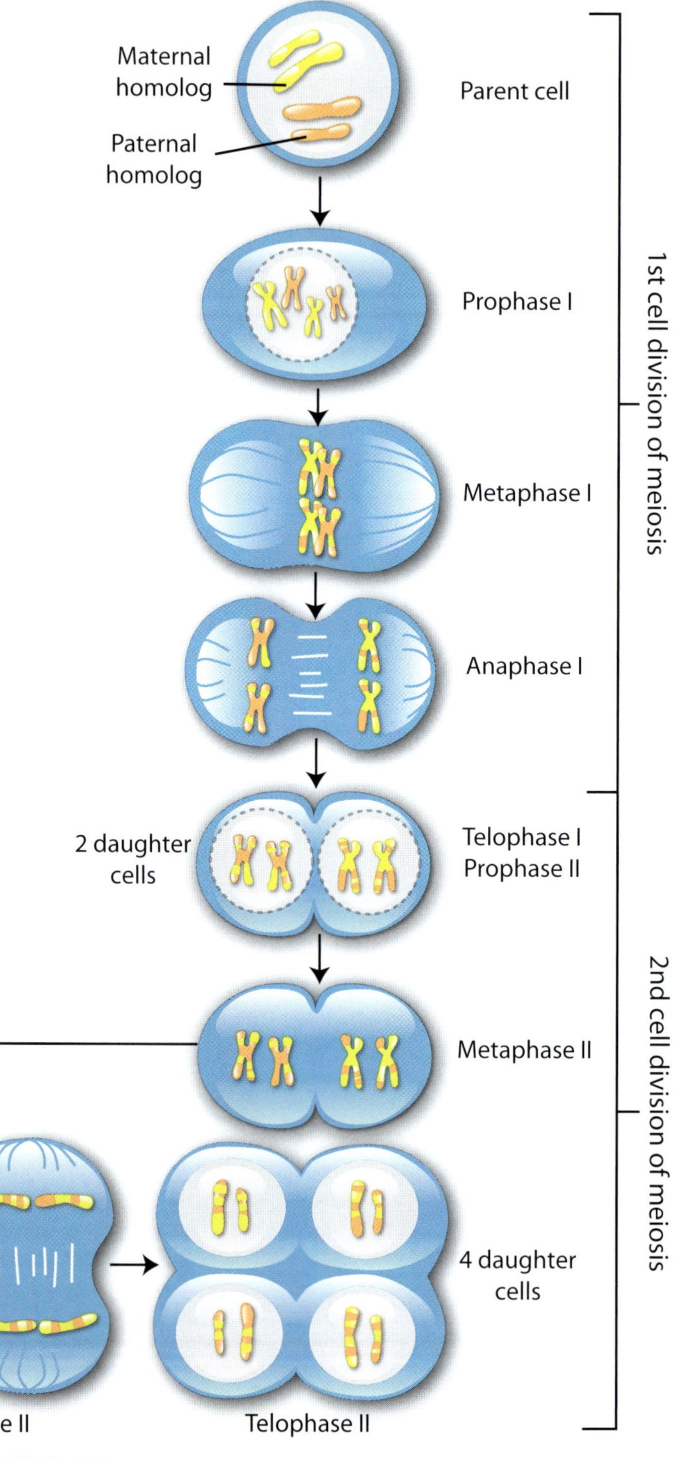

Figure 5.6 The stages of meiosis.

Meiosis I and Meiosis II (Figure 5.6). Listed below are the major phases of each. The key processes of recombination and independent assortment occur in the first two phases of Meiosis I. Meiosis II is very similar to mitosis, but involves the primary concept of genome reduction to create gametes that have only a single genome complement (1N).

Meiosis I

Prophase I: After the DNA has been replicated, the nuclear membrane disappears and the chromosomes condense. The centrioles (protein structures) move to the polar regions of the cell and also form spindle fibers—the structural fiber network that moves the chromosomal material around the cell. This is the first phase of the whole process, but one of the most important. At this stage, the paternal and maternal chromosomes pair up with their respective counterparts (homologs). After they pair up, they exchange sections of their DNA, creating new chromosomes that are a patchwork of both maternal and paternal DNA. This process is called *homologous recombination*.

Metaphase I: The paired-up chromosome homologs line up in the center of the cell along the metaphase plate, or equatorial plane. They become physically attached to each other and the spindle fibers at the chromosome's centromeres. The kinetechore, a protein structure, not only binds to the centromere as an attachment point for the spindle fibers, but also as a feature capable of splitting apart the sister chromatids (DNA copies). In this phase, independent assortment takes place as the chromatids begin to split apart and start the process of moving to one of the two poles.

Anaphase I: The spindle fibers pull the homologous chromosomes made of attached sister chromatids apart from each other to opposite poles in the cell. At this point, the cell elongates in preparation for the first cell division.

Telophase I: The chromosomes arrive at the poles and the cell splits into two daughter cells (cytokinesis), each with a new nucleus that has formed around the newly recombined chromosomes.

Meiosis II

Prophase II: Once again the chromosomes condense, the nuclear membrane disappears, and the centrioles move to the poles and form spindle fibers.

Metaphase II: The chromosomes line up in the center of the cell along a new metaphase plate that is positioned in a 90° offset (perpendicular) compared to the metaphase plate in Meiosis I. Chromosomes have microtubules attached for movement within the cell.

Anaphase II: The chromosomes composed of attached sister chromatids split, are pulled apart, and arrive at the poles via the spindle fibers. The chromosome content at the poles is actually divided into a total of four haploid sets. Remember, haploid (1N) equals one set of chromosomes, which is half the normal genome (2N).

Telophase II: A final round of cell division occurs that produces four daughter cells, each with a haploid genome content. These haploid cells are known as gametes and form sperm cells in males and egg cells in females. When a sperm and an egg cell unite or fuse, the normal diploid (2N) genome content is restored.

The process of meiosis is a highly complex and orchestrated event that involves the action of thousands of interacting genes, functional RNAs, and proteins. Incredibly, it also involves processes that control and govern key randomization events necessary for genetic diversity, adaptation, and the survival of animal and plant life in general.

Once again, we can bring the practices of engineering and computational programming into play as an analogy. Randomization routines under tight computational control are key elements of many complex computational algorithms and computer programs. They do not happen by chance, but involve intricate engineering and logic. The highly designed, engineered, and controlled randomization of fragments of DNA within a linear space of 3 billion nucleotides (human genome) times two (6 billion nucleotides) in an environment as complex as a eukaryotic cell is absolutely incredible!

Evolutionists have not been able to propose any naturalistic explanation of events that could have led to such miraculous biological processes that represent the core essential features of life.

Summary

1. *Mitosis* is the term used to identify the incredibly complex and immeasurably precise processes by which cells divide. *Cell division* is absolutely necessary for an organism to develop, grow, and maintain its functions. There is no plausible explanation for the origin of these processes from an evolutionary perspective. Engineering is so evident and processes too precise to exclude the possibility of an omniscient Creator.
2. The *human cell* has two sets of chromosomes, each with 3 billion base pairs of DNA. These 6 billion bases must be replicated quickly and accurately every time a cell goes through mitosis. This process occurs continuously throughout the human body. Whenever random DNA changes occur, the result is *not* helpful, but instead can promote cancer-like diseases that cause the cells to proliferate out of control. Precise function in DNA replication and mitosis produces order and purpose. Random action produces disorder and death.
3. *Meiosis* occurs during the formation of sperm and egg cells. These processes, while similar to mitosis, require the unique *recombination* of two complete sets of chromosomes into one wholly unique chromosome set. Meiosis is a one-way, two-stage continuous process that only takes place in cells that are responsible for the reproduction of the organism.
4. The process of meiosis is a highly complex event that *orchestrates* the action of thousands of interacting genes, functional RNAs, and proteins. Even *randomization routines* are tightly controlled. These routines are necessary for genetic diversity and adaptation of all human, animal, and plant functions. It is impossible to conceive of purposeless and random chemical processes acting over unimaginable eons of time producing the precision of meiosis.
5. *Randomization routines* are key elements of many complex computational algorithms and computer programs. These routines are executed under tight computational control and require intricate engineering and logic. The controlled randomization of fragments of the DNA within the nucleotides of eukaryotic cells represents an immensely higher level of complexity. The only reason to ignore the obvious design and creation of these unique processes is to reject the Creator Himself.

Brain's Complexity "Is Beyond Anything Imagined"

Brian Thomas, M.S.

The brain has for a long time been compared to man-made computers in its astounding ability to process, store, and route information. But a new imaging technique has revealed that just one brain's connections and capacities far outnumber and outpace those of all the world's computers. And this makes the question of the origin of brains that much more difficult for naturalistic explanations.

The imaging technique, called array tomography, detected light emitted by mouse nervous tissue that had been bioengineered to produce proteins that glow. Additional luminous chemicals were added, and these attached to specific areas in the mouse brains, adding more colors and allowing for the detection of much more information.

New computer software processed all the data to produce stunning three-dimensional images of never-before-seen brain cell connections. A study was published in the journal *Neuron* to showcase the usefulness of array tomography, but the technical paper also provided broad implications for neurobiology.

Array tomography could "resolve fine details at the level of synapses,"[1] allowing researchers to view the highest resolution of detail among nerve images yet. They could see individual synapses, the tiny connection points that link nerve cells together, as well as their different types and subtypes.

"[The researchers] found that the brain's complexity is beyond anything they'd imagined," according to an article appearing in the Health Tech section of the online news source CNET.[2] For instance, they found that the total number of synapses in a brain roughly equaled the number of stars in 1,500 Milky Way galaxies! And memory patterns and tiny on/off switches, which were long thought to reside in the larger neuron cell bodies, were instead found to be smaller than the tiny synapse connections. Each of the neurons imaged in the study serves thousands of synapses.

Stanford University professor and senior study author Stephen Smith said that "one synapse may contain on the order of 1,000 molecular-scale switches. A single human brain has more switches than all the computers and routers and Internet connections on Earth."[2] This research multiplies the brain's overall computing power far beyond what was previously known.

The more complicated a system is, the stronger it argues for having been intentionally designed. And brains certainly qualify, despite assertions that random-acting natural processes somehow assembled them. In these cases, the burden of proof lies heavily on those who insist that such systems are not in fact what they plainly appear to be: the products of intentional ingenious design.

The God of the Bible stands as the most tenable source of the specified complexity of interconnected neurons upon which human and much animal life depends. Until a naturalistic alternative can explain how a self-healing, adaptive, cosmic-sized Internet of connectivity has been shrunk down to the size of a brain, then it is best to identify this hyper-tech design as being the product of a real Designer. And until an objective body of evidence can legitimately debunk the Bible's historicity and proven accuracy, then it is best to identify this Designer as the Creator and Sustainer revealed in Scripture.

References

1. Micheva, K. D., et al. 2010. Single-Synapse Analysis of a Diverse Synapse Population: Proteomic Imaging Methods and Markers. *Neuron*. 68 (4): 639-653.

2. Moore, E. A. Human brain has more switches than all computers on Earth. *CNET News*. Posted on news.cnet.com November 17, 2010, accessed January 6, 2011.

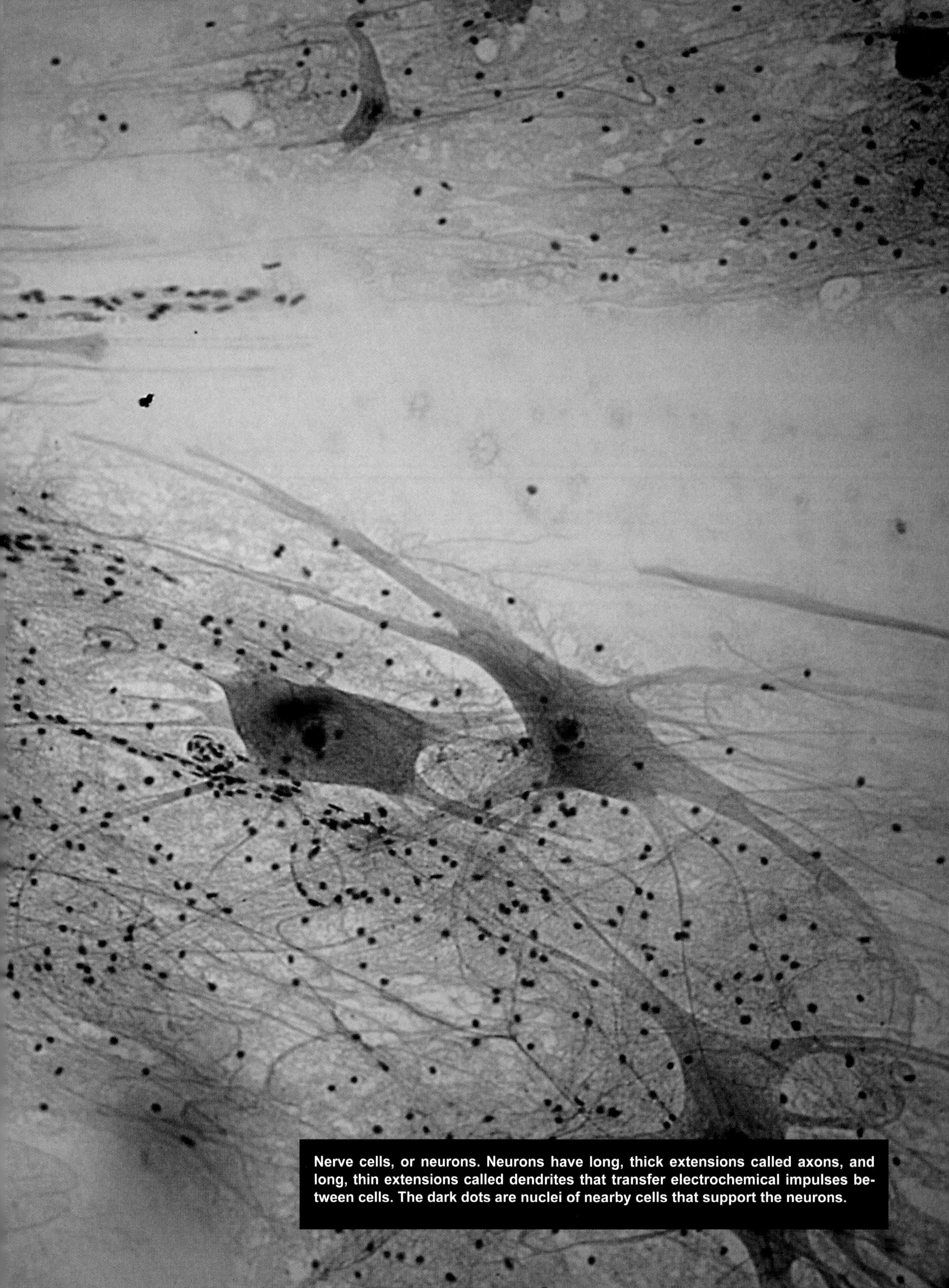

Nerve cells, or neurons. Neurons have long, thick extensions called axons, and long, thin extensions called dendrites that transfer electrochemical impulses between cells. The dark dots are nuclei of nearby cells that support the neurons.

Chapter 6

Cell Signaling: The Miracle of the Biological Network

One of the most important aspects of cellular function is the complex, internetworked communication that occurs via chemical and electrical impulses. The two main types of cell communication—that within the cell and that outside of the cell—are interrelated. Cell signaling can be initiated from outside the cell through cell surface receptors, which are highly specialized protein molecules imbedded in the cell membrane that extend out of the cell into the extracellular matrix (Figure 6.1). The extracellular portion of the cell surface receptor contains binding regions specific for certain signaling molecules (receptor ligands).

Cell Surface Receptors

There are different types of cell receptors and different mechanisms used to transmit and receive signals. *Ion channel-coupled receptors* function through the binding of neurotransmitters, a specific class of signaling molecules. These receptors control the permeability,

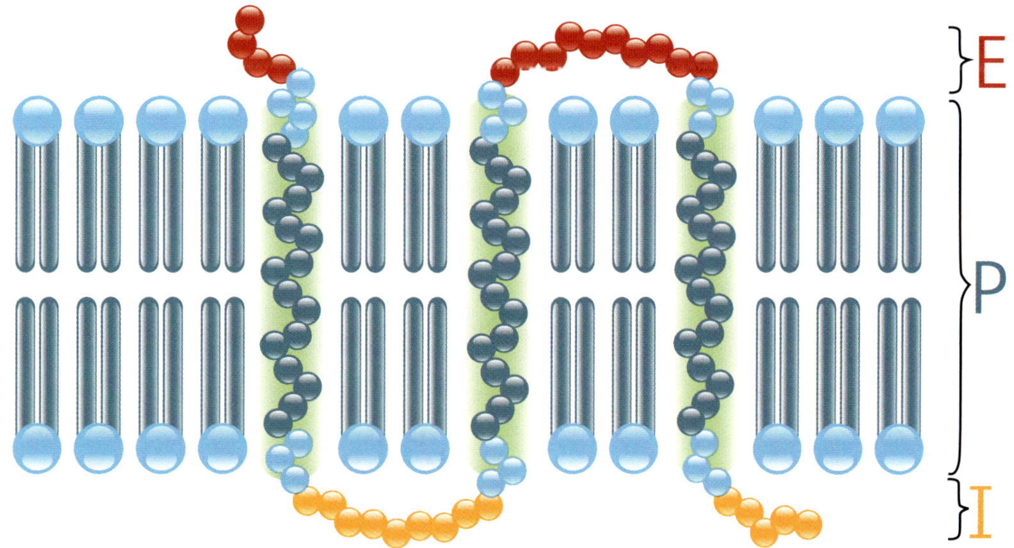

Figure 6.1 Illustration of a transmembrane receptor protein. The E (extracellular) portion protrudes outside the cell, providing a cell surface receptor for extracellular signaling molecules (ligands). The I (intracellular) region protrudes into the cell as part of the signaling cascade to the interior of the cell. The P (plasma membrane) region is the portion imbedded in the cell's phospholipid bilayer plasma membrane. This image is also an example of a multipass receptor protein, which passes (loops) in and out of the membrane several times. A single-pass receptor would only transverse the membrane once.

or openness, of electrochemical signal pathways (ion channels) through the cell membrane. Ion channels are selective for ions with specific electrical charges, such as sodium or potassium molecules that have positive charges. The binding of the cell receptor with the neurotransmitter ligand (e.g., acetylcholine) causes the ion channel to open or close, changing the permeability of the membrane for a particular ion and causing a shift in electrical potentials across the cell membrane. This type of chemical-based electrical action is controlled by the amount and duration of the neurotransmitter released and bound at the receptor. The action of ions crossing the cell membranes causes electrical impulses between cells, creating the nerve impulses needed for muscle movement, thought processes, and sensory processes—making all animal life possible. Amazingly, these processes operate at lightning-fast speeds and provide the basis for all the day-to-day life activities that we take for granted.

Another type of cell surface receptor is called a *G protein-coupled receptor*. These receptors work indirectly in the cell by regulating the activity of a separate plasma membrane-bound protein or ion channel. A G protein (trimeric GTP-binding protein) is used to control this interaction. Guanine triphosphate (GTP) is an energy molecule used in the cell that is similar to the ATP molecule mentioned previously. The G protein receptors are large multipass transmembrane proteins, looping in and out of the cell membrane several times. They act on a variety of proteins in the cell, depending on the G protein's target inside the cell. G protein receptor-mediated signaling occurs through a cascade of events that begins with the activation of the cell receptor by an extracellular signal (ligand). The activated G protein then activates another protein inside the cell or opens up an ion channel to create an electrical potential. If the G protein activates an internal protein, a signaling cascade is initiated that involves the sequential activation of multiple proteins (like a row of dominoes falling). The end result of a signaling cascade in the cell is the triggering of a variety of genes and gene networks in the cell's nucleus (Figure 6.2).

A third type of cell surface receptor is an *enzyme-coupled receptor*. These are single-pass transmembrane proteins that are heterogeneous (diverse) in structure but similar in function. Like the other receptor proteins embedded in the cell membrane, they have an extracellular ligand-binding site and an intracellular catalytic or enzyme-binding site in the cytoplasm. They can actually be an enzyme, or they may associate directly with an enzyme that they activate. Most enzyme-coupled receptors are protein kinases or associate with protein kinases, which are a diverse group of regulatory proteins that add phosphate (PO_4) groups to other proteins, causing them to become activated in most cases. These consecutive phosphorylation events are key features of an intracellular signaling cascade.

Because many intracellular signaling proteins act like molecular switches, a mechanism must exist that enables them to turn other proteins on and off. This is accomplished within the cell through two methods: phosphorylation and GTP binding. When a receptor protein is bound by its ligand (an extracellular signal molecule), a protein kinase is activated (directly or indirectly), which in turn phosphorylates (adds phosphate groups to) another protein in the signaling cascade. This protein may be another kinase or a different type of enzyme, but in either case, it is activated through this phosphorylation event.

There are two types of protein kinases: Serine-threonine protein kinases, the most abundant, which phosphorylate serine and threonine amino acids; and tyrosine kinases, which phosphorylate tyrosine amino acids. These activated proteins can be turned off by phosphatases, which are enzymes that remove the activating phosphates. The activity of many proteins in the cell depends on the activity of its associated kinase and phosphatase.

The second type of molecular switch within the cell is

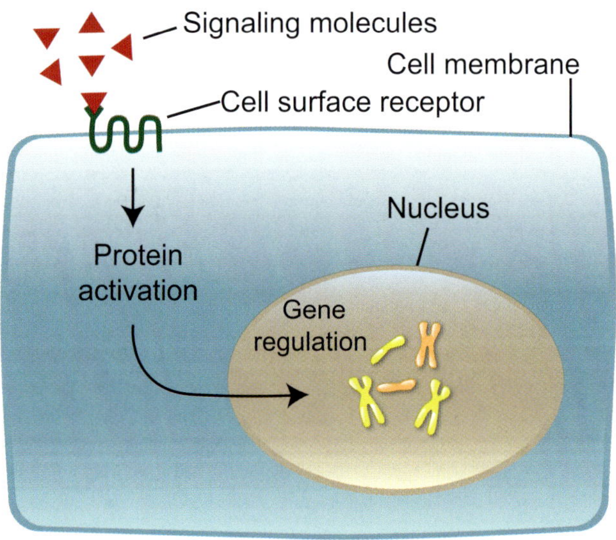

Figure 6.2 Diagram of a signaling cascade.

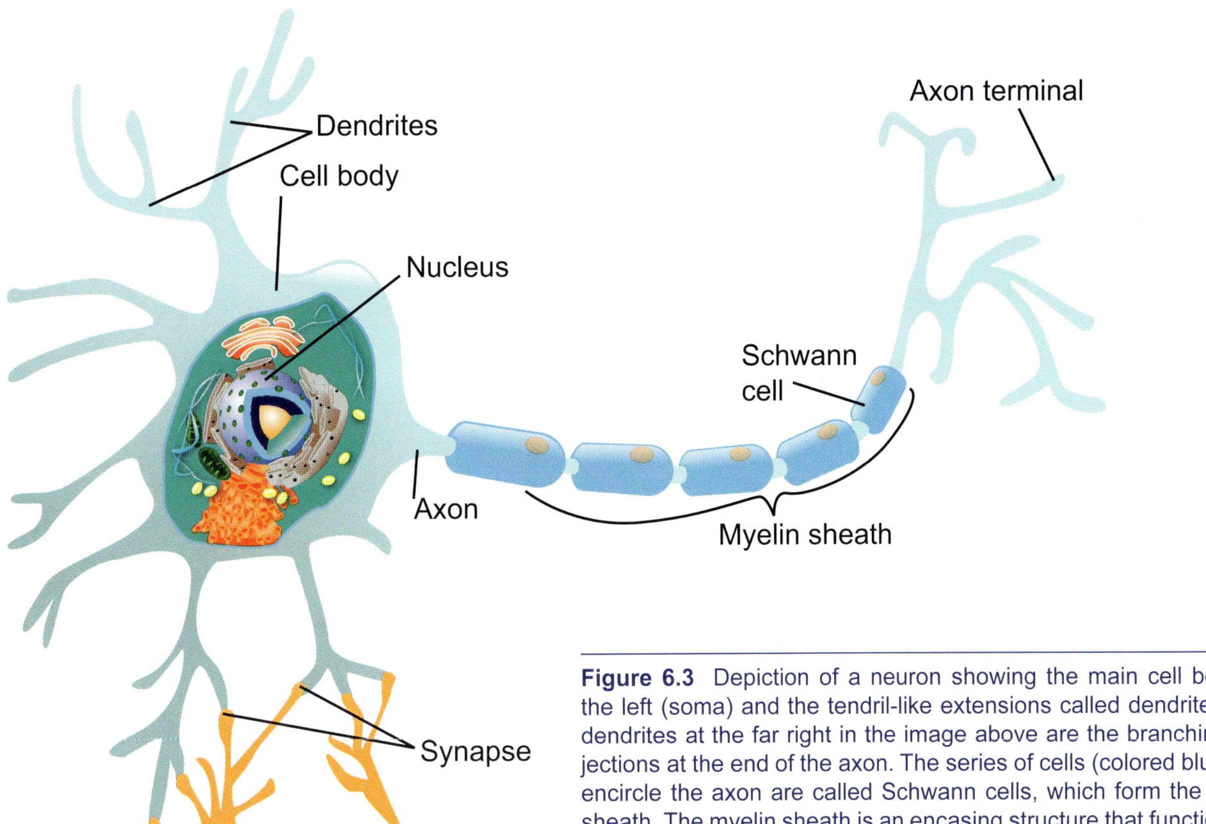

Figure 6.3 Depiction of a neuron showing the main cell body on the left (soma) and the tendril-like extensions called dendrites. The dendrites at the far right in the image above are the branching projections at the end of the axon. The series of cells (colored blue) that encircle the axon are called Schwann cells, which form the myelin sheath. The myelin sheath is an encasing structure that functions as a type of electrical insulation for the axon.

based on GTP-binding activity. When a GTP molecule is bound to a protein, it becomes active in a fashion similar to the phosphatase/kinase activity mentioned above. When a GDP (guanine diphosphate, or GTP minus one phosphate) is bound, the enzyme becomes inactive. Some GTP-binding proteins also have this ability built in so they can shut themselves off.

The last type of receptor to mention is an enzyme-coupled receptor called an RTK (receptor tyrosine kinase). When an extracellular signal binds to the RTK (a single-pass transmembrane alpha helix), portions of two RTKs associate with each other, bringing their kinase domains together. This allows the two protein chains to cross-phosphorylate each other on multiple tyrosines. Once these sites are phosphorylated, the kinase activity of the receptor increases. Other regions of RTKs serve as docking ports for specific intracellular signals. Docking specificity is maintained when intracellular signals recognize the phosphate and the surrounding features in the protein. Once the protein signal molecule is bound to the RTK, it may be phosphorylated and activated, or the binding alone may induce a shape change that activates the signal and the pathway. The RTKs can bind different combinations of signaling proteins and have the capacity to activate many different types of cellular responses.

Neurons

In vertebrate animals and humans, neurons (also known as nerve cells) make up the tissues of the brain and spinal cord, which together are referred to as the central nervous system. Neurons are highly specialized cells that conduct electrochemical nerve impulses in the nervous system, which is a highly complex, yet incredibly efficient, network that extends from the spinal cord to virtually every location in the body. Remarkably, just the human brain and spinal cord contain approximately 100 billion neurons. The neuron is the core cell type responsible for virtually every action, whether voluntary or involuntary, that the human body makes.

Even though it is somewhat variable in structure depending on its location, the neuron utilizes a fairly consistent design pattern. As shown in Figure 6.3, its major features include a soma (main cell body), axon (tendril-like extension of the soma), and dendrites (branching projections). In addition, there are a series of cells that encircle the axon called Schwann cells that form the myelin sheath, an encasing structure that functions as a type of electrical insulation for the axon. Schwann cells are essentially glial cells that support both myelinated and un-myelinated neurons (i.e., both those with and without a myelin sheath). This structure and configuration enable systemwide completion of extremely complex tasks in just fractions of a second. An electrical charge is created in the main body of the neuron that translates to a chemical-based signal that jumps a gap and is converted into an electrical signal upon its reception at a connecting neuron.

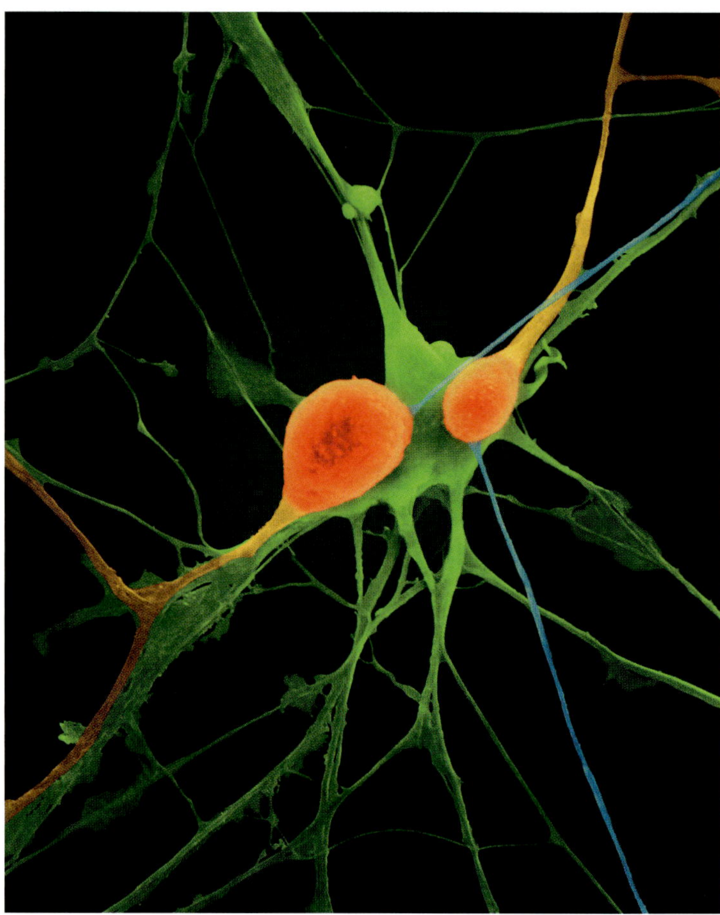

Figure 6.4 Light microscope picture of a human neuron cell.

Neurons are quite amenable to study in the lab and can be dyed to provide contrast in light microscope photography (Figure 6.4). These cells can also be cultured in petri dishes and other containers. This enables specific studies to be conducted, such as observing the effects on the cells of various chemicals and toxins that are common in our food and environment.

There are several different types of neurons throughout the human body that vary in function, size, and mechanisms of signaling. The interaction between nerve cells, which occurs through the signaling cascade described above, is facilitated by gaps between ganglia, or groups of nerve cell bodies. These gaps are called *synapses* and act as the signal transmission zones between cells. The cell extensions in these gap areas contain a wide variety of cell receptors. Many of these cell receptors serve as drug targets for pharmaceutical products (e.g., analgesics and psychiatric products). The space between the receptors is a complicated yet controlled mix of chemical messengers that regulate many types of bioelectrical activity. Once again, there is virtually no way to devise an evolutionary model that could produce the immense interdependent complexity of the internetworked neural systems that exist in animal life.

The above material serves as an introduction; we have barely scratched the surface of the complexity that is actually present. Entire books could be written on the subject and still not adequately explain the complexity of the nervous system. Truly, the presence of incredibly complex nervous systems is another clear and resounding argument for creation.

Summary

1. *Cell signaling* both from outside and inside the cell is absolutely necessary to maintain life. These complex network communications occur continuously through both chemical-based and electrical impulses. They permit reflex reactions (jerking away from pain, for instance), as well as controlled movement (sitting, standing, walking, talking, singing, and a host of other selected activities). Without these lightning-like signals within and between the cells, the whole organism would soon cease to exist.
2. *Cell surface receptors* work through different mechanisms. Some use ion channels that select for specific electrical charges (like sodium or potassium molecules with positive charges), while others produce a cascade of events triggered by certain kinds of proteins. Still others react to enzyme-coupled receptors that add phosphate groups to the chain, activating or stopping the signals. These all have specialized functions and targeted responses that never react randomly, but are triggered and work as needed. The internetworked complexity and relationship of cell signaling throughout the entire organism can only be defined as miraculous.
3. *Neurons* (nerve cells) are highly specialized cells that are present in vertebrate animals and humans. Their job is to conduct electrochemical impulses through an incredibly complex and exceedingly efficient network that extends from the brain through the spinal cord to every location in the body. Neurons contain a main cell body, tendril-like extensions, and branching projections. Scientists believe that the brain and the spinal cord contain some 100 billion neurons that are interconnected and work in concert. This unique cell system enables virtually instantaneous completion of complex tasks throughout the entire network.
4. *Synapses* are the connections through which the information to and from the neurons is passed. Scientists believe that each neuron connects through approximately 10,000 synapses. This incredible network of information storage, retrieval, and transmission enables motion, sight, hearing, smell, and taste, as well as the ability to reason, think, dream, plan, remember, and anything else done with the brain. Scientists have determined that the number of synapses in a human brain exceeds the number of the stars in the Milky Way by 1,500 times!
5. Computer information (coded messages) is transferred over the Internet in "packets"—short bursts of data that are "driven" along networks at electronic speeds, then reassembled and decoded so that the information is legible on the receiving computer screen. The neuron and synapses of the nervous system can be compared to that process, even though nothing that human beings have been able to manufacture comes close to the observable operation of the human brain. The probable amount of information data that flows among neurons and through the synapses in the central nervous system exceeds the supposed 300 sextillion stars in the universe! It is hard to imagine such complexity arising by chance.

Human Reproduction

Randy J. Guliuzza, P.E., M.D.

A new life is started the moment a human sperm cell unites with a human egg. Sounds simple, doesn't it? A person can decide for himself, but he will need to follow along very closely to catch all of the details and carefully piece them together, just as he would follow a skillful mystery.

The voyage of a single sperm cell from production to fertilization begins with rapidly dividing cells within a testis called spermatogonia. These divisions are crucial to place in the sperm cell 23 chromosomes—exactly half the number within normal human cells. When the sperm fuses with the egg, which also has undergone divisions within the mother's ovary, the full complement of 46 chromosomes will be present. However, cellular mechanisms allow slight variations in the information contained on certain portions of the chromosomes to be shuffled during the divisions. This feature ensures that each sperm and egg carries the correct information to make a normal human, but each is different as to the exact traits that will be expressed by the new person. The genetic combination in the newly fertilized egg will be totally unlike that of any person who has ever lived before or ever will be born afterwards—truly resulting in an absolutely unique individual.

The sperm starts out as a round, immobile cell. It is surrounded by other cells in the testis called Sertoli cells, which function only to transform the sperm cell into a lean swimming machine that is capable of carrying its genetic cargo to the egg. Sertoli cells transfer nutrients to the developing sperm from the blood stream, since at this point in development the sperm must not be in contact with blood. Large amounts of cellular fluid within the sperm, called cytoplasm, are also removed by the Sertoli cells, and internal cellular components are precisely rearranged so that the sperm begins to take on the shape of a long and slender cell with a whip-like tail. An important structure, the "acrosomal body," that will eventually develop highly erosive enzymes—able to dissolve the membranes around other cells—is made by Sertoli cells at the newly-developed head of the sperm and sealed in a protective coat.

A high concentration of the male hormone testosterone in the testes is essential to make normal sperm. Where does it come from? Far from the testes, the brain's hypothalamus will release "gonadotropin-releasing hormone," which stimulates the pituitary gland to release "follicle-stimulating hormone" and "lutenizing hormone." These make their way via the blood stream to the testes. Lutenizing hormone stimulates other cells in the testes called "Leydig cells" to manufacture prodigious amounts of testosterone. Follicle-stimulating hormone now causes the Sertoli cells to produce "androgen-binding proteins" that will bind the testosterone produced in Leydig cells and concentrate it inside, where it will have its effect on the developing sperm. As the testosterone level increases, it also circulates throughout the body. When the correct concentration of testosterone (along with a concentration of the hormone "inhibin," which is made in Sertoli cells) circulates back to the hypothalamus and pituitary gland of the brain, these structures are signaled to stop secreting their hormones. Without this stimulus, Leydig cells decrease production of testosterone until the circulating concentration drops to a level that will trigger the cycle to start all over again—keeping it in perfect balance.

Recall how the sperm are being kept from contact with the blood. They are locked behind very tight junctions between Sertoli cells that make a collective configuration called the "blood-testis barrier." Why? A male does not begin producing sperm until puberty, and the markers on the new sperm cells have not been programmed into his immune system. The male's immune system is programmed to recognize specific combinations of protein markers on the outside of his cells as belonging to his own body—but that programming takes place while he is still in his mother's womb. Were it not for this barrier, sperm cells would be recognized as foreign cells by the male's own immune system and destroyed, rendering the male sterile. If the junction between Sertoli cells is broken, such as what happens when the testes become inflamed during an infection with Mumps virus, antibodies can make their way from the blood stream past the barrier and destroy the developing sperm.

Sperm placed inside a woman find themselves in a very hostile environment, with features that either destroy microscopic entities or block entrance into her body. The normal vaginal environment is very acidic (pH 3.5), which suppresses dangerous bacterial overgrowth but also kills sperm. Fluids produced by the male seminal vesicles are part of the semen and temporarily neutralize (pH 7.5) the acid. The neutral environment then activates the sperm. A thick sticky mucus plug also blocks the small cervical opening into the uterus. However, another product of semen called prostaglandins causes this mucus to become more liquid-like. Not coincidentally, the mucus may also have been made even thinner by an estrogen surge in the woman around the time she ovulates an egg. Now sperm are able to swim through the mucus into the uterus—all the while converting substances in the mucus to energy.

The uterus is protected by millions of cells of the woman's immune system that kill microscopic invaders. This obstacle is overcome by substances in the semen that have local, but very broad spectrum, immunosuppressive effects that blunt her immune response in the area of the semen. This may leave the woman vulnerable to infection, but another substance in semen, "seminalplasmin," can kill bacteria and has a protective effect. Normally, coordinated movements of mobile hair-like projections called cilia on some cells lining the uterus, coupled with slight rhythmic contractions of the uterus, produce a defensive fluid current that pushes things out of the uterus—which would be impossible for the sperm to swim against. Yet another product of semen after making contact with the uterus causes these coordinated actions of the woman's uterus to reverse direction and pull the semen and sperm up into the uterus and assist the sperm on its journey.

Surprisingly, freshly deposited sperm are incapable of fertilizing an egg. Many features of the sperm are changed by substances that are made within the female reproductive tract. Remember the sperm's acrosomal body discussed earlier? One of the most important changes, known as "capacitation," is when uterine secretions remove glycoproteins from the protective coat of the acrosome. This allows the erosive enzymes from many sperm (after contacting the egg) to break down a protective coat of cells around the egg and expose its cell membrane so that yet another sperm can make its way to the egg for fertilization. This elaborate coordination between female secretions and male sperm is protective for the male, since without the protective coat around the acrosome, high concentrations of sperm in a man's body could destroy the function of his reproductive organs if the erosive enzymes were released prematurely.

The acrosome is coated with the protein "bindin" that will adhere only to special species-specific receptors on the egg, ensuring that only sperm from the same species can fertilize the egg. In less than a second after the sperm's contact, many channels in the egg's membrane open, allowing an inrush of positively-charged sodium ions. This creates an electrical charge across the outer surface of the egg that blocks other sperm from fertilizing it and inactivates all remaining "bindin" receptors on the egg. Concurrently, substances just inside the egg's cell membrane are released that bind up water molecules and cause the membrane to swell up to permanently detach any remaining sperm on the outside.

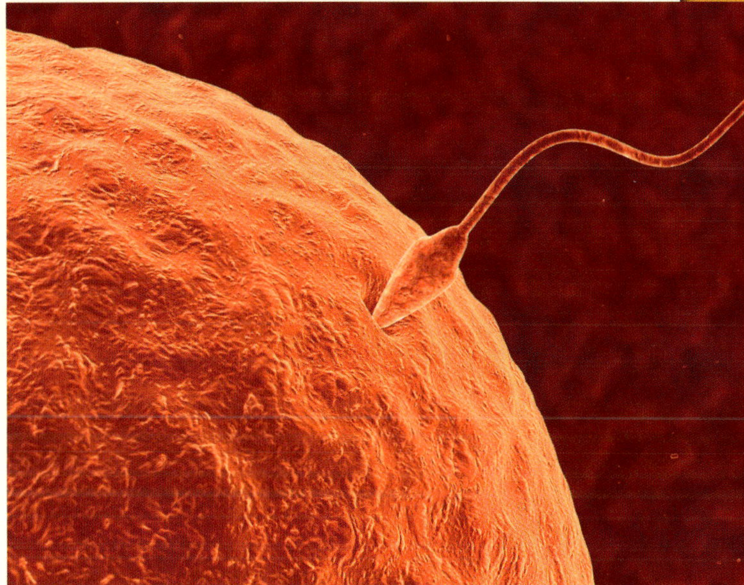

These blocks prevent entrance of genetic material from any other sperm into the egg, which would be fatal to baby and may also be to mother. Once united, tube-like structures in the egg rapidly build and then project from the egg and pull the nucleus of the sperm into the egg—the first cell of a new person.

Amazing? Actually, the detail could go far beyond this simple description. As seen, the level of coordinated interaction to get any viable offspring exceeds the cellular level, extends past the reproductive system, pulls in the neurologic, hormonal, and circulatory systems, and demands substances that are produced independently by the male to modify the actions of the female body or the materials made by her—and vice versa. Evolutionary literature is rife with speculative stories about the origination of these processes, but devoid of any real scientific evidence to explain them. The only viable explanation is that these processes were placed by the Lord Jesus in the first parents, Adam and Eve, fully functional right from the beginning.

Muscle cells, or fibers. Striated muscle fibers are very long and tightly grouped. Each fiber is densely packed with contractile proteins that produce physical motion of body parts when innervated, such as when the esophagus contracts during swallowing.

Chapter 7

The Cytoskeleton and the Extracellular Matrix: How Biology Achieves Shape, Form, and Movement

Cells have an internal network of specialized rod-like molecules. This network, or *cytoskeleton*, resembles the internal skeletal and muscular structure that provides form and function to the human body and to animals in general. The cell's structure and related functions are primarily dependent on two protein molecules—*actin* and *tubulin*—whose subunits assemble into rod-like structures called cytoskeletal filaments. Both of these play important roles in the spatial organization, structure, and mechanical properties of the cell. Actin helps determine cell shape and is required for cell locomotion and movement. Tubulin, present in structures called *microtubules*, helps determine the positions of organelles in the cell matrix and plays a key role in directing intracellular transport (Figure 7.1).

Actin Filaments and Microtubules

In the cell, actin is present as strands or rods (called actin filaments) that are composed of two-stranded helical polymers made of actin monomers, or individual protein units. These filaments possess a polarity. The plus end is more dynamic and is typically the site of filament growth where

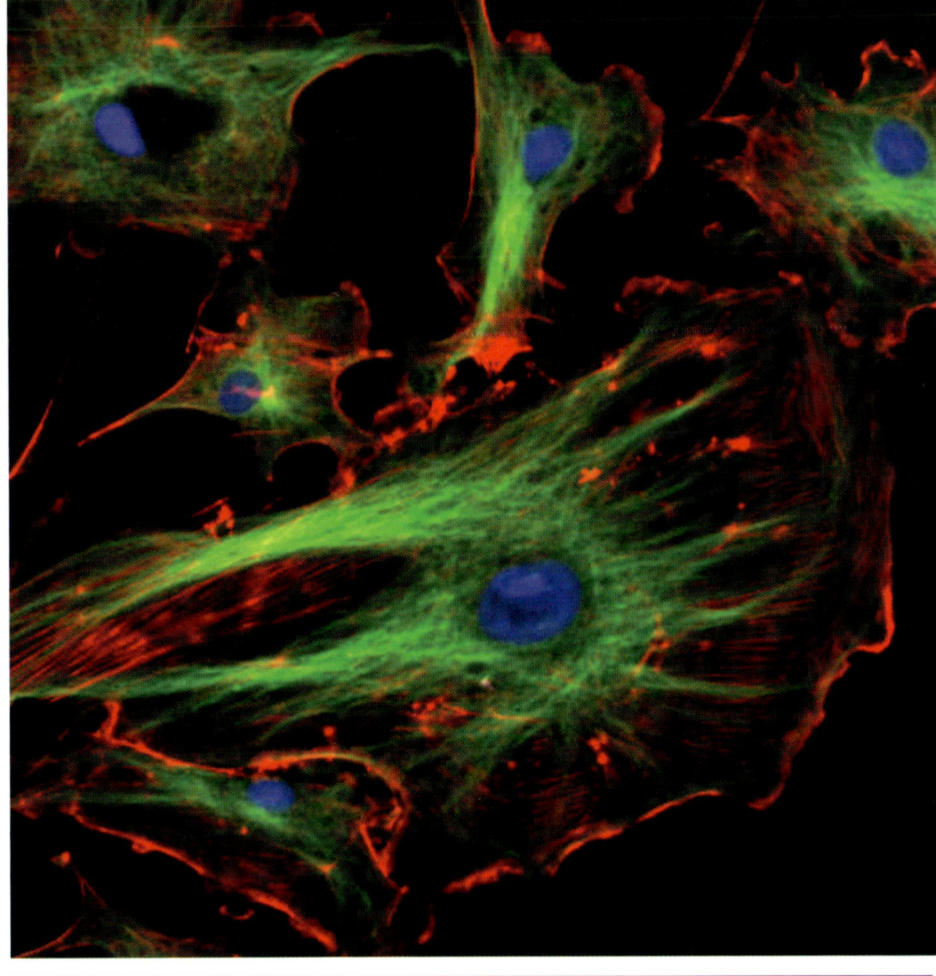

Figure 7.1 Tubulin molecules illuminated via green fluorescence in the single-cell eukaryote *Tetrahymena*.

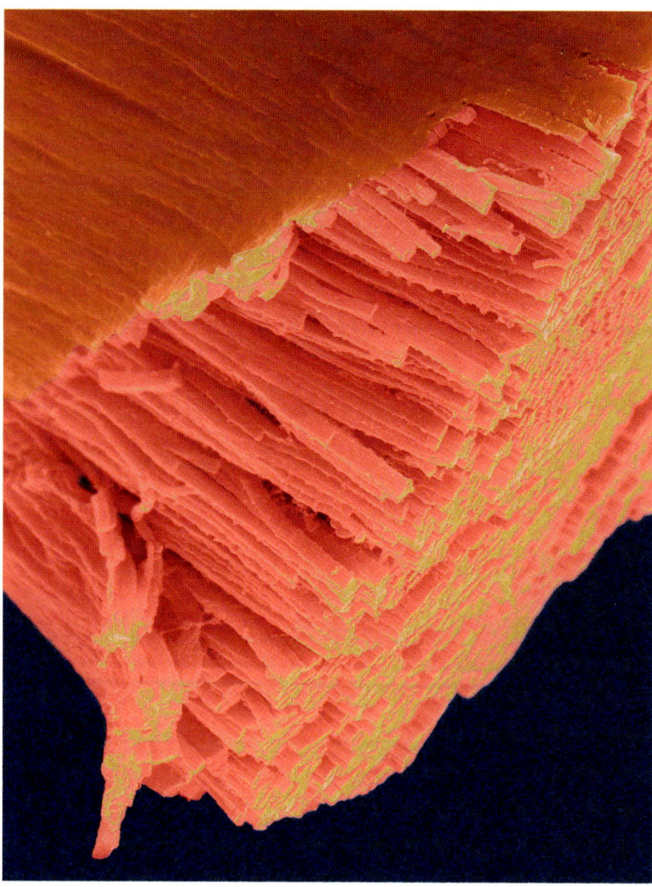

Figure 7.2 Electron micrograph image of actin filaments in mouse embryo. Actin organized as strands or rods called actin filaments are composed of two-stranded helical polymers made of actin monomers (individual protein units).

protein units are added. The minus end is the site of monomer loss where individual protein units are removed. The assembly of proteins at one end of an actin filament and the disassembly of subunits at the other end can occur in concert to cause a pulling motion. This process occurs very quickly and plays an important role in not only moving things around inside the cell, but also in overall cell movement, as in muscle tissues. Actin monomers have an ATP-binding site because the process of assembly and disassembly requires energy input. The whole actin filament is composed of two actin protofilaments that twist around each other in a right-handed helix. They are fairly flexible, but they are usually found in the cell assembled into large-scale structures that provide shape and strength to the cell (Figure 7.2).

The two most common types of tubulin are alpha (α) and beta (β). A tubulin subunit is made of one of each type. Both types have a bound GTP, but in the α-tubulin the GTP is bound so tightly that it never leaves, and thus forms an integral part of the protein. The β-tubulin GTP, however, may be hydrolyzed, which is a chemical reaction that splits a substance into two parts by adding a water molecule. This is important to the structure of microtubules, because GTP is often hydrolyzed shortly after incorporation into the filament. Following hydrolysis, the resulting GDP will remain bound and, for a short time, the filament may exist in the T form (with GTP bound) and in the D form (with GDP bound). The same process happens with actin, but ATP is bound and ADP is left on the older filaments.

The tubulin types α and β come together to form protofilaments (Figure 7.3). Thirteen of these protofilaments, in parallel, form a microtubule. Because the α and β tubulin proteins alternate, the microtubule has a polarity, with α-tubulins exposed on one end and β-tubulins exposed on the other. The crosswise and lengthwise interactions between these heterodimers (two different proteins complexed together) hold the microtubules together. Microtubules are incredibly important to the movement of organelles such as the mitochondria and chloroplast in the cell. They are also an integral component of the processes used during mitosis. They attach to the newly replicated chromosomes and separate the two new sets from each other into new cells.

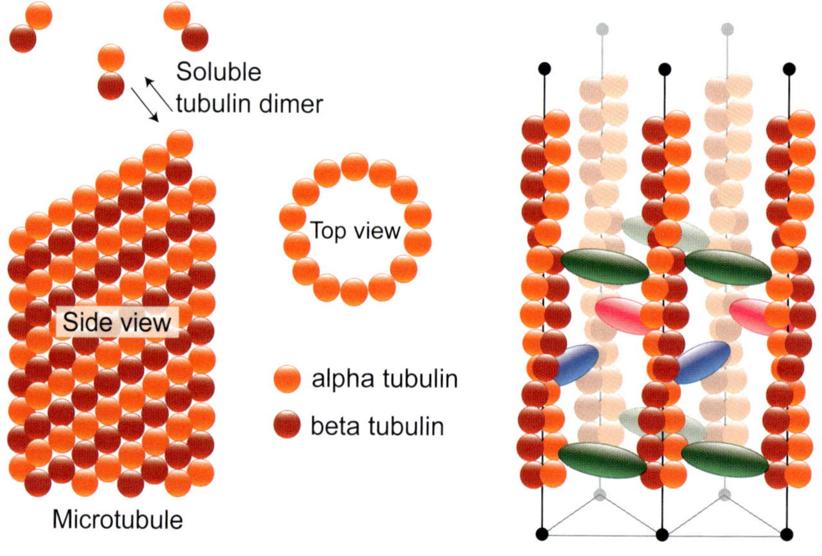

Figure 7.3 Simplified diagrams for the molecular structure of tubulin on the left and actin on the right.

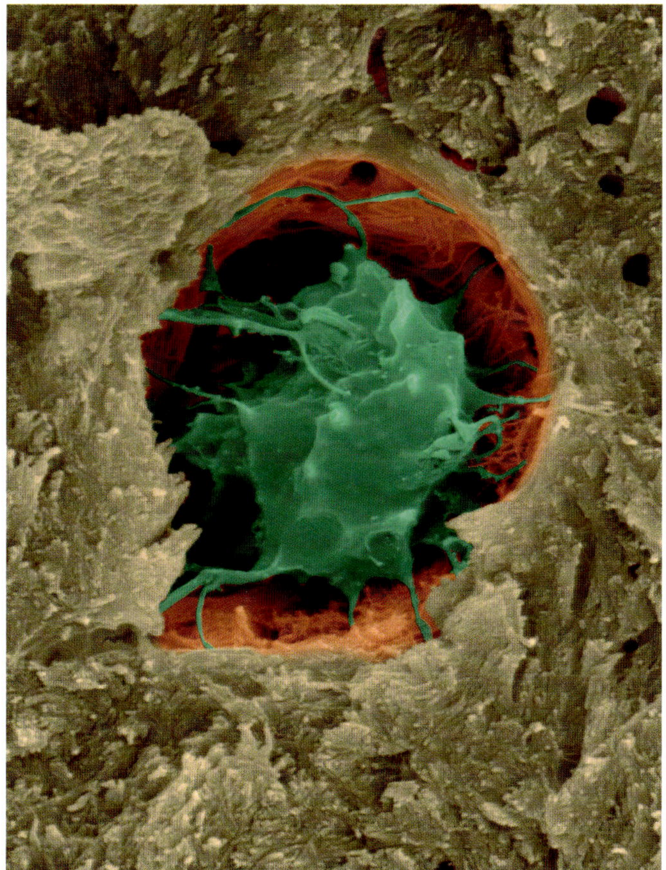

Figure 7.4 The filopodia are long, finger-like extensions that protrude from the lamellipodia framework.

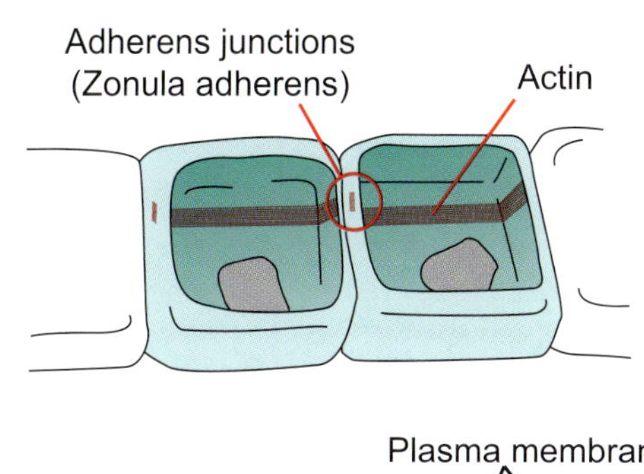

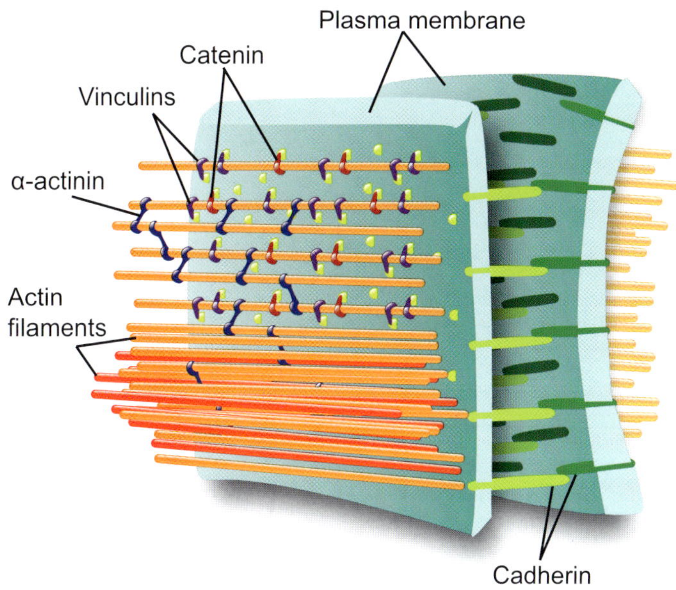

Figure 7.5 Depiction of an anchoring junction via adherens that creates a cell-to-cell adhesion.

Actin filaments are highly concentrated just below the plasma membrane because of their involvement in the movement of cells. This activity is facilitated by extensions of the cell called *lamellipodia* and *filopodia*. The lamellipodia are actin-filled bulges along the mobile section of a cell that work in a wave-like manner, creating locomotion. The filopodia are long, finger-like extensions, also actin-filled, that protrude from the lamellipodia. Figure 7.4 shows these structures and how they relate to each other. These extensions function in cell locomotion, neuron formation and function, bone growth and maintenance, and many other activities specific to the cell type and its role in the body. The actin filaments are also very important in muscle-type tissue. There, they form "stress fibers" that have extreme contractile properties and exert high levels of tension.

One of the major differences between microtubules and actin filaments is how they nucleate themselves (i.e., how they start the formation of fibers). For microtubules, a unique kind of tubulin, γ-tubulin, is involved. At the microtubule-organizing center in the cell, a γ-tubulin ring complex is found. This ring complex binds two proteins that subsequently bind to α and β tubulins, and the whole complex forms a template for the 13 protofilaments of the microtubule. For actin fibers, nucleation occurs most often at the site of the plasma membrane and is catalyzed by the ARP complex and a group of proteins called *formins*.

This ARP complex is made of actin-related proteins and nucleates actin filament growth from the minus end, so that the plus end rapidly elongates. In other words, the ARP complex holds on to the minus end, allowing actin monomers (single actin proteins) to attach to the plus end and extend the fiber. Formins nucleate the assembly of actin filaments by capturing two actin monomers. Unlike the γ-tubulin and the ARP complex, however, the formin proteins do not stay associated with the minus end, but move along with the growing activity at the plus end.

Junctions between Cells

Junctions or connections between cells are very important for holding cells together in groups, for cell-to-cell communication between neighboring cells, and for providing shape and form to organs and tissues. This is especially important considering that the organs in the body are formed from large groups of cells that must function in unison.

The first type of junction is an *anchoring junction* (Figure 7.5). This includes both cell-to-cell adhesions (joining two cells together) and cell-matrix adhesions (joining a cell to its surrounding material). The anchoring junctions transmit stresses between cells because of the attachments of cytoskeletal filaments in the cell. In cells where actin is the anchoring filament, the cell-to-cell junctions are called *adherens* and actin-linked adhesions form the cell-matrix junctions. In cells where intermediate filaments provide the anchoring mechanism, the junctions are called *desmosomes* (cell-cell junctions) and *hemidesmosomes* (cell-matrix junctions).

Occluding junctions seal gaps between cells in epithelial tissues, such as surface areas like skin or the inner lining of the intestine. This enables epithelial cells to form an impenetrable or semi-permeable barrier. Occluding junctions also allow cells to maintain their sense of orientation and polarity.

Channel-forming junctions—also known as *gap junctions* in animals or *plasmodesmata* in plants—create passageways between the cytoplasm of neighboring cells. The cytoplasm of attached cells in a group or an organ is continuous, and the associated cells are able to maintain the same general chemistry. These types of junctions, however, typically do not allow large macromolecules like proteins to pass between cells.

Signal-relaying junctions are another important category. A good example of a signal-relaying junction is the synapse, a junction between nerve cells. In cells of the central nervous system, signals transfer from one cell to another across connected membranes at the synapse (specialized sites of porous contact). Other important signal-relaying junctions are the immunological synapses of the immune system and the transmembrane ligand-receptor signaling contacts that facilitate a wide variety of cell-to-cell communication. Gap junctions and plasmodesmata are both channel-forming, signal-relaying junctions, allowing the passage of small molecules between the cytoplasm of adjacent cells.

The gap junctions found in animal cells are made by the channel-forming proteins in the plasma membrane that were discussed in Chapter 6. These channels connect neighboring cells both electrically and metabolically, and permit inorganic ions and water-soluble molecules to freely pass between cells. The actual structure and shape of each gap junction is dependent on the properties of the structural proteins and cell-signaling proteins involved. These proteins determine the permeability and interactive characteristics of the gap

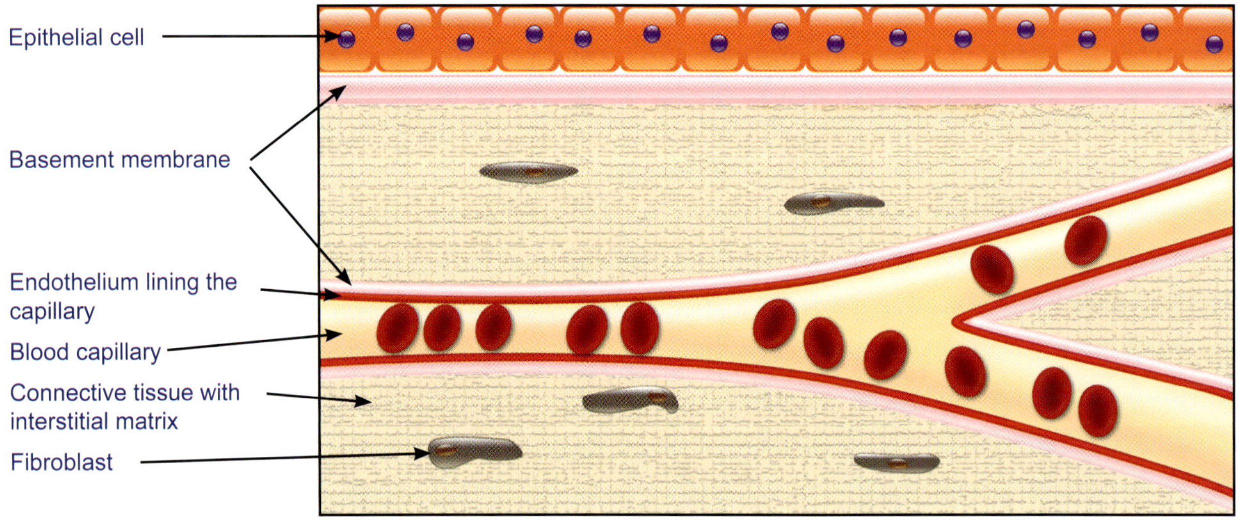

Figure 7.6 The extracellular matrix depicted in the context of human skin (the epidermis).

junction. Gap junctions are able to change their permeability depending on changes in pH (acidity), concentration of Ca2+ (calcium ions), or binding of extracellular signals to cell surface receptors. Gap junctions provide a variety of critical functions to cells, including the transfer of action potentials in electrically communicative cells, coordination of physiological activities among cells, and the coordination of development and growth in embryos.

Plasmodesmata are the channels or pores between cells in plants that are formed by a shared continuous plasma membrane between adjacent cells, rather than channel-forming proteins like in the gap junctions found in animals. In plants, a cell wall barrier between neighboring cells is an added structural feature that largely eliminates the possibility of gap junctions. Most plasmodesmata also have a specialized portal called a *desmotubule* running through the middle of them. This is a narrow cylinder that makes the endoplasmic reticulum contiguous among connected cells, facilitating cell signaling. The end result is a signaling "hotline" between the cell's command centers, their nuclei.

Instead of being inserted into the plasma membrane like the channel proteins that form gap junctions in animals, plasmodesmata are formed during the process of cell division (cytokinesis). They can also be inserted "after the fact" in a *de novo* fashion into pre-existing cell walls. Plamodesmata are extremely versatile and can allow the passage of many different types of mol ecules, including DNA regulatory proteins, mRNAs, and various small molecules.

The Extracellular Matrix

The extracellular matrix is a critical feature in every multicellular organism (Figure 7.6). It provides a binding mechanism, a protective structure, and a mechanical substrate for the associated cells. The extracellular matrix is made up of many proteins, which are integrated into a hydrated gel composed of glycosaminoglycan chains. The glycosaminoglycans are linked to a core protein to form the larger proteoglycan molecule structures needed. Because the proteoglycan molecules are negatively charged, they attract positively charged small molecules. This causes water to be drawn into the matrix and forms the chemical basis for hydration, the absorption and retention of water. The water pressure in the matrix associated with hydration provides the ability to tolerate compressive forces, giving organs and tissues the feature of rigidity. Interestingly, the proteoglycans not only provide important structural traits in the extracellular matrix, but their protein subunits also perform many important functions related to signaling.

The main cell-based component of the extracellular matrix is a class of cells called *fibroblasts*. These are specialized cells that help to orient the polarity of the matrix. They also produce the macromolecules that form the large proteoglycan structures. The cytoskeleton within fibroblasts controls the orientation of the matrix outside of the cell. Studies in animal tissues suggest that fibroblasts work through the matrix to also create the tendons, ligaments, and sheathes of connective tissue around organs.

Another type of protein complex present in the extracellular matrix is *elastin*, a molecule that provides elasticity. Elastin proteins attach together to form an extensive network of fibers and sheets that have the ability to stretch and recoil. Elastin is also very hydrophobic, meaning that it repels water and forms the selective water barriers needed by all types of multicellular organisms.

The extracellular matrix is also composed of fibrous proteins, like collagen. Collagen can form larger complexes that are composed of three individual collagen chains wrapped around each other. Type I collagen molecules are assembled into larger structures called collagen fibrils. These associate a second time into larger bundles called collagen fibers. These larger fibers are used as building blocks that come together and form different types of structural associations, allowing form and function to exist at the whole organism level. Collagen provides the critical strength and resilience that the matrix must provide for large-scale skeletal features. It also provides binding sites for cells, tissues, and organs.

Interestingly, collagen is present in virtually all higher animals that require muscular movement for locomotion on land and in aquatic environments. Like bone tissue, collagen is very durable and easily preserved in fossils. Collagen tissue is often excavated by paleontologists in conjunction with bone tissue from both aquatic and terrestrial extinct creatures.

In fact, these collagen proteins found in fossils are in such pristine condition that in many cases they can be subjected to protein sequencing technologies. In recent years, several groups of evolutionary scientists have published collagen protein sequences derived from fossils of both *Tyrannosaurus rex* and *Brachylophosaurus canadensis* (duck-billed hadrosaur). These are dinosaurs that supposedly lived and died more than 50 million years ago.

The big issue is that proteins should not be present if fossils are as old as evolution claims. These discoveries indicate that fossils are not millions of years old, but only several thousand years old—which is exactly what would be expected, based on the biblical history of a young earth.

Summary

1. *Cytoskeleton* networks are an internal network of rod-like molecules that provide form and function to multicell organisms. They are particularly active in providing shape and movement to the organism. Two protein molecules, *actin* and *tubulin*, play important roles in controlling the mechanical properties of cells.
2. *Actin filaments* are constructed in a two-stranded helical rod that can be assembled at one end and disassembled at the other in such a way to cause a pulling motion—very important for muscle tissues. These twisted filaments are usually assembled into large-scale structures that give shape and strength to the cell. Actin-filled bulges (called *lamellipodia*) along the mobile section of cells work together in waves, providing cell locomotion, neuron formation, bone growth, and other activities specific to the needs of the cells and their function in the body.
3. *Tubulin protofilaments* are bound together in groups of 13 to form a *microtubule*. These microtubules allow the movement of organelles such as the mitochondria and the chloroplast and are active in mitosis and meiosis. The formation of the tubulin fibers is facilitated by a unique *gamma* tubulin that forms a ring complex of two other proteins which, in turn, bind the *alpha* and *beta* tubulins together. That newly formed set becomes the template from which the 13 protofilaments are developed. Microtubules are then held together by the special properties of the alpha and beta types of tubulin.
4. Several types of *cell junctions* are necessary. Large groups of cells must work together in concert to provide shape and form to organs and tissues. Essentially, junctions must connect one cell to another cell and also make cell connections to the surrounding matrix. *Occluding junctions* form barriers, *channel-forming junctions* create passageways between cells, and *signal-relaying junctions* facilitate a wide variety of cell-to-cell communications.
5. The *extracellular matrix* is made up of many proteins. These include classes of cells known as *fibroblasts* (creating tendons, ligaments, and connective tissues), *elastin* (forming networks of fibers that stretch and recoil), and *collagen* (providing strength for large-scale skeletal features). These matrices are present in animals and humans, and are absolutely necessary for the organism to function. Once again, the intricate complexity of these individual functions, as well as the integration of these functions within the whole, are so clearly designed that it seems incredible that any thinking person could view them as having been randomly produced over eons. The Bible indicates that the Creator is a God of order, purpose, and planning, and the observable universe verifies that even the "invisible" side of God's wonderful nature and His "eternal power" are "clearly seen" by the creation that He has brought into existence (Romans 1:20).

The Mysteries of Stunning Soft Tissue Fossil Finds

Brian Thomas, M.S.

The controversial soft tissue finds of North Carolina State University paleontologist Mary Schweitzer are gaining renown, and for good reason. She found organic material in fossilized dinosaurs and other creatures that should not have existed after having supposedly been buried for millions of years. These groundbreaking discoveries are fascinating, and they raise big questions about the standard understanding of fossils and geology.

In 1992, while peering through a microscope at dinosaur bone fragments that had their hard minerals removed, Schweitzer discovered red discs that appeared identical to reptile red blood cells. She wrote in the December 2010 issue of *Scientific American*:

> And those ruby microscopic structures appeared only in the vessel channels, never in the surrounding bone or in the sediments adjacent to the bones, just as should be true of blood cells.[1]

The find was an utter shock because of the long-held belief that fossilized bones have had all their original materials replaced by external minerals. Thus, a colleague's admonition to prove they *aren't* red blood cells has motivated Schweitzer's research ever since. So far, she has been unable to do this. She has also been unable to prove that fresh-looking blood vessels, various soft protein samples, and structures identical to bone cells (osteocytes) are anything other than what they appear to be.

These discoveries presented profound mysteries. Researchers had always expected that "after millions of years, buried in sediments and exposed to geochemical conditions that varied over time, what was preserved in these bones might bear little chemical resemblance to what was there when the dinosaur was alive."[1] But this is an extraordinary understatement. In fact, given what is known of tissue and biomolecular decay, after all that time there should be nothing preserved at all and, thus, *zero* resemblance to any original dinosaur tissue.

Schweitzer summarized the conundrum that her work presented, stating:

> Our findings challenged everything scientists thought they knew about the breakdown of cells and molecules. Test-tube studies of organic molecules indicated that proteins should not persist more than a million years or so; DNA had an even shorter life span.[1]

Actually, proteins take much less time than a million years to spontaneously decay.[2] But neither ten thousand nor one million years are anywhere near the millions of years needed to make standard geological long-age scenarios plausible.

Thus, fossil soft tissue is caught between a rock and a hard place. On the one hand, it should not be there at all, based on lab studies. On the other hand, its presence is obvious.

Therefore, to accommodate soft cells from fossils into the standard long-ages view, scientists have had to make a tough choice. One option has been for biochemists to deny that the paleontologists are actually seeing what they claim is evident. Another choice has been for paleontologists to deny what the biochemists' lab experiments have measured.

Schweitzer appears to have favored the second option. She asked, "Why are these materials preserved when all our models say they should be degraded?"[1] In this case, modifiable "models" have been substituted for easily repeatable, experimentally derived "data" in a classic case of bait-and-switch.

A third choice has been to play the role of the ostrich with its head buried in the sand and ignore the entire discussion. But there is a fourth option: Once the presumed need for "millions of years" is removed from age estimates, all the data harmonize.

That would also mean that the presence of still-soft original tissues buried in rock layers coincides with biblical records about the timing of earth history, which indicate that all of life was created only thousands of years ago.

References

1. Schweitzer, M. H. 2010. Blood from Stone: How Fossils Can Preserve Soft Tissue. *Scientific American.* 303 (6): 62-69.
2. Thomas, B. How Long Can Cartilage Last? *ICR News.* Posted on icr.org October 29, 2010, accessed December 14, 2010.

Bacterial cells of the genus *Staphylococcus*, or “staph.” Although most staph species are innocuous and are normally found on skin, some varieties cause disease—especially when they cluster within mucus shields that protect against the host’s immune tactics.

Chapter 8

Cells in Creation, Sin, and Redemption

Plants

According to the book of Genesis, cells (biological life) were involved in the creation events on Days Three through Six of the creation week (Figure 8.1 illustrates the sequence of creation). On Day Three, God created the earth's plants—terrestrial and, although not explicitly stated, presumably aquatic plants as well.

> **Day Three:** And God said, Let the earth bring forth grass, the herb yielding seed, and the fruit tree yielding fruit after his kind, whose seed is in itself, upon the earth: and it was so. And the earth brought forth grass, and herb yielding seed after his kind, and the tree yielding fruit, whose seed was in itself, after his kind: and God saw that it was good. And the evening and the morning were the third day. (Genesis 1:11-13)

It may have been more orderly and optimal to create plants prior to animal life for a number of reasons. First, plants are the foundational component of the biosphere, providing atmospheric oxygen for terrestrial animal life and scrubbing the atmosphere of carbon dioxide through *photosynthesis*. In addition, plants also provide food and shelter to animal life all over the planet. While these points are only speculative, creating animal life on an earth void of plants would have been a non-optimal situation.

Take, for an example, the preparation of an aquarium. The water, rocks, sand, aquatic plants, filter system, etc., are prepared prior to adding fish and other aquatic creatures. In fact, even after putting the aquarium environment together, the whole system is also typically allowed to equilibrate for at least several hours prior to adding the fish. While this is a somewhat crude and oversimplified analogy to the creation of the earth and

Figure 8.1 A graphical depiction of the days of creation.

Figure 8.2 Photosynthesis enables plants to capture and process energy from the sun.

its life forms, it helps to illustrate the point that the whole creation sequence was done in an orderly pattern that fits well with what we know about plant and animal ecology.

Photosynthesis and Respiration

At the beginning of the creation week, God created light and divided it from the darkness (Genesis 1:3-5). Perhaps this was done as a means of establishing the importance of the 24-hour demarcation, or the solar day as we know it. Obviously this enforces the biblical text regarding the creation week as a literal seven-day week. However, the primary source of energy for photosynthesis in plants is solar radiation, and the sun was not created until Day Four.

> **Day Four:** And God said, Let there be lights in the firmament of the heaven to divide the day from the night; and let them be for signs, and for seasons, and for days, and years: and let them be for lights in the firmament of the heaven to give light upon the earth: and it was so. And God made two great lights; the greater light to rule the day, and the lesser light to rule the night: he made the stars also. And God set them in the firmament of the heaven to give light upon the earth, and to rule over the day and over the night, and to divide the light from the darkness: and God saw that it was good. And the evening and the morning were the fourth day. (Genesis 1:14-19)

Even if no light source was present for the first day that plants existed, it does not challenge the validity of the literal 24-hour day model because plants can easily survive without sunlight for 24 hours or more. Interestingly, most seed-borne, tuberous, or rhizome-propagated plants in the normal cycle of life begin their lives in relative total darkness prior to their emergence from the soil. Prior to this, and for a short period of time thereafter, cellular growth and development are fueled by protein and carbohydrate reserves. Following emergence, the young plant makes a metabolic conversion from stored energy to carbohydrates produced from the activity of photosynthesis (Figure 8.2).

One interesting feature of Day Four from a cell biology perspective is that God takes an apparent 24-hour break in creating multicellular organisms. On this day, the solar energy from the newly created sun was made available to the world's plants. So why do we have this one-day buffer between animals and plants?

Once again, we can only speculate using what we know about cell biology. Perhaps a one-day break between the creation of plants and animal life allowed for an equilibration or stabilization of the earth's atmosphere. The two key chemical substrates in the photosynthetic reaction driven by solar energy are carbon dioxide and water. The byproduct of this reaction is oxygen (O_2), which is the molecule required for respiration, the metabolic foundation of all biological life. The physiological phenomenon of respiration is a metabolic process whereby living organisms break down food molecules with the help of atmospheric oxygen.

Like animals, plants use atmospheric oxygen for cell respiration (metabolism), which largely occurs at night. Most of the cell growth in plants happens in the dark when they utilize the carbohydrates produced from photosynthesis during the day as an energy source. It is highly conceivable that the atmosphere on the first day following the creation of plants may have been carbon dioxide-rich and oxygen-deficient, potentially toxic to animal life; therefore, God did not create the animals on the fourth day. As a result, the atmosphere was properly equilibrated in regard to the carbon dioxide and oxygen ratios so that God could proceed with the next logical step in creation—animal life. Of course, this is only speculation, because God is omnipotent and could have prepared the atmosphere for animal life in an instant.

The six-day creation sequence is perfectly logical, given our understanding of cell biology. We have only recently (in the past 50 years) discovered and researched the biochemical pathways associated with respiration and photosynthesis. These cell processes validate the importance of the creation order within the framework of a literal 24-hour day. The original author of the Genesis account and the early Bible translators most likely did not know these scientific details. The chaotic nature of non-biblical creation legends around the world do not account for this type of logic and scientific detail. Therefore, the logical scientific progression of the creation account further confirms that the creation week based on a literal 24-hour day is a powerful apologetic argument for biblical authority.

Water, Air, and Land Creatures

Some Christians attempt to develop a "biblical" model of creation in which the days of the Genesis creation week correspond to long, vast ages of geological time (the day-age theory). But what we know of cell and ecosystem biology presents enormous challenges to this theory. One big problem is that plants and animals have a vital symbiotic relationship. Plants rely on animals for many facets of reproduction, seed dispersal, and other diverse aspects. Plants could not survive without animals for very long, and vice versa.

The worldwide creation of plants on Day Three and a day of global photosynthesis on Day Four prepared the earth and its atmosphere for the creation of animal life on Day Five.

> **Day Five:** And God said, Let the waters bring forth abundantly the moving creature that hath life, and fowl that may fly above the earth in the open firmament of heaven. And God created great whales, and every living creature that moveth, which the waters brought forth abundantly, after their kind, and every winged fowl after his kind: and God saw that it was good. And God blessed them, saying, Be fruitful, and multiply, and fill the waters in the seas, and let fowl multiply in the earth. And the evening and the morning were the fifth day. (Genesis 1:20-23)
>
> **Day Six:** And God said, Let the earth bring forth the living creature after his kind, cattle, and creeping thing, and beast of the earth after his kind: and it was so. And God made the beast of the earth after his kind, and cattle after their kind, and every thing that creepeth upon the earth after his kind: and God saw that it was good. And God said, Let us make man in our image, after our likeness: and let them have dominion over the fish of the sea, and over the fowl of the air, and over the cattle, and over all the earth, and over every creeping thing that creepeth upon the earth. So God created man in his own image, in the image of God created he him; male and female created he them. And God blessed them, and God said unto them, Be fruitful, and multiply, and replenish the earth, and subdue it: and have dominion over the fish of the sea, and over the fowl of the air, and over every living thing that moveth upon the earth. And God said, Behold, I have given you every herb bearing seed, which is upon the face of all the earth, and every tree, in the which is the fruit of a tree yielding seed; to you it shall be for meat. And to every beast of the earth, and to every fowl of the air, and to every thing that creepeth upon the earth, wherein there is life, I have given every green herb for meat: and it was so. And God saw every thing that he had made, and, behold, it was very good. And the evening and the morning were the sixth day. (Genesis 1:24-31)
>
> **Day Seven:** And on the seventh day God ended his work which he had made; and he rested on the seventh day from all his work which he had made. And God blessed the seventh day, and sanctified it: because that in it he had rested from all his work which God created and made. (Genesis 2:2-3)

Following the 24-hour break in the creation of major cell-based life, aquatic animals and land creatures capable of flight were made on the fifth day. Interestingly, the other forms of terrestrial animal life were not created until the sixth day, along with the first humans (Adam and Eve), the climax of creation.

After each major creation event (plants, aquatic creatures, flying creatures, and animals), it is acknowledged that the creation is good. The Genesis text repeatedly states that "God saw that it was good."

Consequences of the Curse

However, things today are obviously not good. So what happened? This is the question raised in everyone's mind in light of the corruptness of mankind and the out-of-balance and stressed state of nature. Questions even arise from the perspective of cell biology, such as why there are so many problems for humans, plants, and animals like diseases, cancer, and various single-cell pathogens.

The fact of the matter is that even cell biology exists in a corrupted state as a result of the Fall of man in the Garden of Eden, caused by Adam and Eve's rebellion against God (Genesis 3). At its core, the Fall shows the root of sin—man's attempt to become like the Creator via the pathway of secret or forbidden knowledge. There is only one omnipotent and omnipresent Creator God, and everything else has been "created," including man. When man disobeyed God in his misguided and ill-conceived quest to be like Him, a curse was placed on all creation. Rebellion is never committed in a vacuum. This act of disobedience has had consequences not only for every generation of man since, but also for all of creation, including cell life. Although cells are clear and resounding testaments to an infinitely mighty Creator God, they also testify of man's original and ongoing rebellion against God.

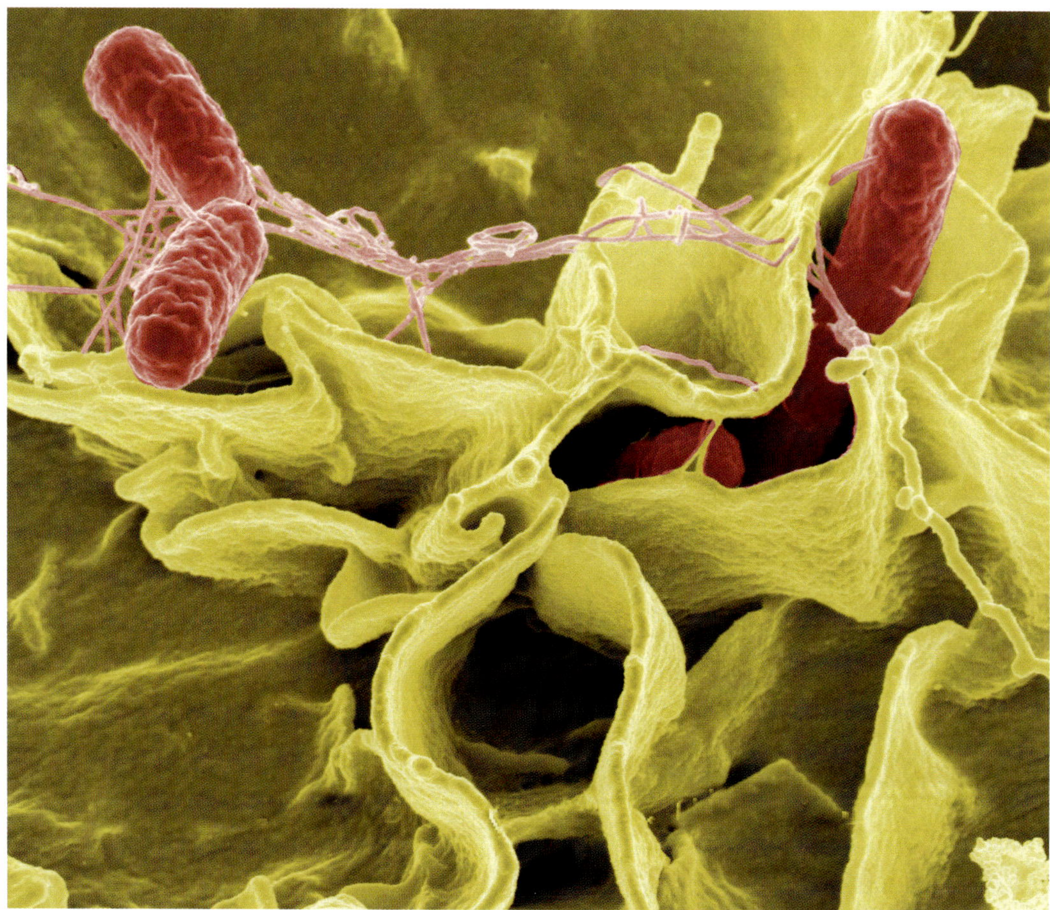
Figure 8.3 Pathogenic *Salmonella* bacteria invading human cells grown in culture.

One of the most often-voiced arguments against the presence of a good and merciful Creator God is the presence of cell-based diseases and pathogens (Figure 8.3). In regard to biologically related cell diseases, if it were not for the multitude of built-in, cell-based protective measures created by God (e.g., the body's innate immune system), all life would have quickly been extinguished long ago. However, mankind is both personally (e.g., via smoking, substance abuse, etc.) and corporately (industry and business-related toxin deployments in our environment) in rebellion against the innate defense systems God created and placed in our cells.

In regard to the presence of diverse pathogens and parasites that seem to have no other purpose than to cause disease and raise havoc, the fact of the matter is that in virtually all cases examined so far, a factor related to the "cursed earth" state has been altered in the organism's environment or genetic makeup to cause it to change its food source, life cycle, natural environmental surroundings, and/or host/symbiotic relationships. In the course of one or more of these factors being changed, the organism has become pathogenic and harmful in some way. When evaluating these parasites and pathogens within this cursed-earth paradigm, it is necessary to do so on a case-by-case basis, which has been successfully accomplished by creation scientists for a range of organisms.

The apostle Paul best addressed this whole issue in Romans 8. The good news is that this current cursed-earth state of nature will not last forever—deliverance from this corruption is on the way.

> For the earnest expectation of the creature waiteth for the manifestation of the sons of God. For the creature was made subject to vanity, not willingly, but by reason of him who hath subjected the same in hope, because the creature itself also shall be delivered from the bondage of corruption into the glorious liberty of the children of God. For we know that the whole creation groaneth and travaileth in pain together until now. (Romans 8:19-22)

Summary

1. The creation sequence was done in an orderly pattern that fits well with what is known about plant and animal ecology. For example, the creation of plants on Day Three provided the atmospheric conditions and food needed for the animals created on Days Five and Six.
2. God gave plants the ability to process energy from the sun through *photosynthesis*. The creation of the sun on Day Four provided a one-day break in the creation of multicellular organisms. This might have been done to allow plants to balance the carbon dioxide and oxygen in the atmosphere before animals arrived.
3. Some people try to stretch the days of creation into long "ages" of geological time (the day-age theory), but this won't work because plants and animals have a vital symbiotic relationship. They couldn't have existed without each other for long periods of time.
4. Adam and Eve's rebellion against God in the Garden of Eden resulted in a curse on all of creation, including cell life. This has led to diseases and illness-causing germs and parasites.
5. The pathogens and parasites that cause disease and damage were not originally created to be that way. Due to the curse of sin, their original functions were changed through alterations in their environments or genetic makeup. But the earth's cursed state will not last forever—God will one day deliver it and His people from corruption.

Immune Systems, The Body's Security Force

Randy J. Guliuzza, P.E., M.D.

Good neighborhoods provide families a lot of protection, but even the best of communities remain vulnerable to the threat of criminals invading their homes. Our human bodies are also vulnerable to foreign invaders such as bacteria, viruses, fungi, and parasites. But when these infection-causing microbes break in where they don't belong, they face a serious defense force, eventually to be caught and destroyed by a highly trained, cell-sized army equipped with a sophisticated array of weaponry. That security force is called the human immune system. Designed with amazingly dynamic communication networks that pass information back and forth between hundreds of millions of cells, the human immune system strategically fights off microscopic invaders and remembers them each time they attack the body.

Without this defense system, none of us could survive, much less reach maturity. When the immune system becomes weak due to disease, immaturity, or deterioration, death is far more likely from infection or cancer. Medical intervention attempts to keep the immune system intact, but little help can be offered once it collapses.

Preparing the Immune System for Conflict

Like any conflict, the first priority must be to distinguish friend ("self") from foe ("non-self"). Cell surfaces are covered with hundreds of protein markers differing in type and combination. One very important marker, the MHC1, is on almost every cell (except red blood cells). Like all proteins, DNA specifies its makeup. But MHC markers are special. The possible genetic combinations as to exactly how it will "look" are so large that no two people have the same combination (except possibly identical twins). As cells recycle materials, this marker actually takes tiny portions of their unique cell fragments and displays these on the surface. In the womb, the immune system is "programmed" to recognize "self" by learning, in a sense, what these surface markers look like. Immune cells programmed to recognize self reside in the lymphatic and blood systems. Their color is grayish white not red, so they are called the white blood cells, or WBCs.

New generations of WBCs start forming in bone marrow. Specific hormones in the marrow or the thymus gland control programming so knowledge of "self" is passed on. This is important, because WBCs operate on only one creed—any marker on any surface that is not totally "self" is "non-self" and will be attacked and destroyed. Markers play a pivotal role throughout the process. Non-self obviously includes bacteria, viruses, fungi, parasites, and toxins like snake venom. One's own cells may be included if infected inside by viruses or internally transformed by cancer. If so, portions of these non-self proteins are carried from inside and placed on the cell surface by the MHC. This, in effect, marks its own cell for destruction.

How Microbial Threats Are Identified

Hundreds of millions of microscopic invaders that trigger defensive immune responses are collectively called "antigens." Among the WBCs, "T cells" and "B cells" are the key players in mounting very specific responses to threats and then keeping these in memory. Once these cells mature, they not only recognize self but will express a receptor that will bind primarily to one type of antigen. Each of these cells may have over 200,000 receptors on its surface. The process controlling maturation—colloquially called "cellular boot camp"—is exquisitely selective. If T or B cells fail to recognize their specific antigen or show any reactivity against "self," they are tagged with proteins prompting self-destruction. About two percent of manufactured cells make it through boot camp. These cells have not yet been exposed to antigens so they are "naïve" to their threat. They are sent to lymph nodes, other tissues, or simply circulate waiting to be exposed. Exposure for some is almost immediate, while others may never be exposed.

Since T or B cells make well over one billion different receptors, some researchers assert that no environmental threat exists for which a matching receptor has not already been made. So humans are not passively waiting for invading microbial antigens to determine which receptors will be made. Rather, the body has cells ready to meet any antigen that arrives. Some cells even have receptors for antigens that have never existed in nature. This incredible inventory of receptors is specified by genes in DNA, yet the total number of genes currently identified in humans is about 23,000. Actually, T or B cells in the boot camp only use a few hundred genetic bits of information, but these are controlled by an elaborate process that shuffles segments, allowing an astronomical number of combinations. One receptor region determines its classification and allows it to interact with other cells. Another region is variable and will "fit" to some antigenic marker that initiates identification and the attack.

Microbes commonly invade through the mouth, lungs, or broken skin. Anticipating the invasion are millions of cells called APCs (antigen presenting cell), which are strategically located in places like the tonsils or just under the skin. APCs constantly sample anything entering the body, and invaders are killed, their cells are broken apart, and many tiny portions—the antigens—are carried by these APCs to lymph nodes and "presented" to T or B cells. T cells that "fit" the antigen become activated. Some T cells may also run into APCs on chance encounters while circulating in blood. A race begins the moment an invader enters the body. Many microbes reproduce rapidly, so the body must destroy them while they are fewest in number.

Activating a Specific Immune Response

Activated T cells initiate immediate and wide-ranging responses. At least 20 different chemical signals are sent throughout the body. Some attract "helper" T cells, "natural killer" cells, "cytotoxic" T cells, and other WBCs to home in at the site of infection. B cells are activated by these T cells, APCs, or directly by antigens. The chemical signals also drive many other important functions.

Activated B cells start reproducing lineages of exact clones. Most clones are directed to undergo dramatic internal restructuring, thereby becoming cells able to make an antibody—an extremely important protein for fighting infection. These antibodies have regions that "fit" to antigens via similar genetic shuffling mechanisms as its "ancestor" B cell. However, an intricate cyclical process ensures that the antigen-antibody fit gets progressively better by becoming even more specific. How? The B cell's genes prescribing variable regions have genetically unstable areas guided by a mechanism that fosters hypermutation—totally random genetic mutations resulting in very slight changes to variable regions from one lineage of antibodies to another. A few new antibodies will fit to the antigen a little better than others. Right on schedule, another type of cell takes the best fitting antibody-antigen package to a special location inside lymph nodes. The B cell making that antibody continues to the next cycle, while all other B cells that made less specific antibodies are sent a signal to self-destruct. After about six days of these mutation-selection cycles, thousands of cells will be making highly specific antibodies—and each of these cells manufactures antibodies at rates of about 2,000 per second.

Eliminating Threats

Invaders now face an extraordinarily fierce two-pronged attack. Antibodies may directly neutralize some threats like toxins, cause some antigens to clump together harmlessly, or most importantly, cover other microbes with thousands of antibody "bull's eyes" targeting them for certain destruction. Once microbes are marked as non-self, there are very few places to hide. Once found, immune proteins may attach to microbes and punch hundreds of holes in them, additional cells may split them apart, other cells may attach to them and inject substances causing them to self-destruct, and many more may engulf them and force them into internal "sacks" filled with acid, hydrogen peroxide, or bleach.

Toxins like those from tetanus bacteria or snake bites act so fast that ready-made antibodies from another person or a horse must be injected in victims to neutralize the toxin. More often, after someone's initial exposure to an antigen, a few B cells that made the most specific antibodies transform into long-lived cells called "memory" B cells. When memory B cells (and memory T cells) see that antigen later, the entire immune response revs up stronger and produces huge quantities of antibodies in just a couple of days. Vaccines are compounds containing portions of dead or weakened antigens to deliberately expose people and, thereby, get their immune system primed. Often-fatal childhood diseases like smallpox, measles, and polio have been greatly reduced with this strategy. A newborn's immune system needs developmental time, but is conferred good initial protection by antibodies from his mother that cross the placenta and from his mother's milk.

Conclusion

The immune system doesn't have even one central dedicated organ—it is a functional system dependent on every other body system, which themselves would not survive without immunal protection. It's a complex system, and every area fights selflessly to rid the body of its foe and cleanse it of deteriorated cells. The immune system's Creator, the Lord Jesus, is even more astounding. For "Christ also loved the church, and gave himself for it" (Ephesians 5:25). As typified in the immune cells He created, He, too, is mighty to save and protect His Body, even when acting to His own detriment.

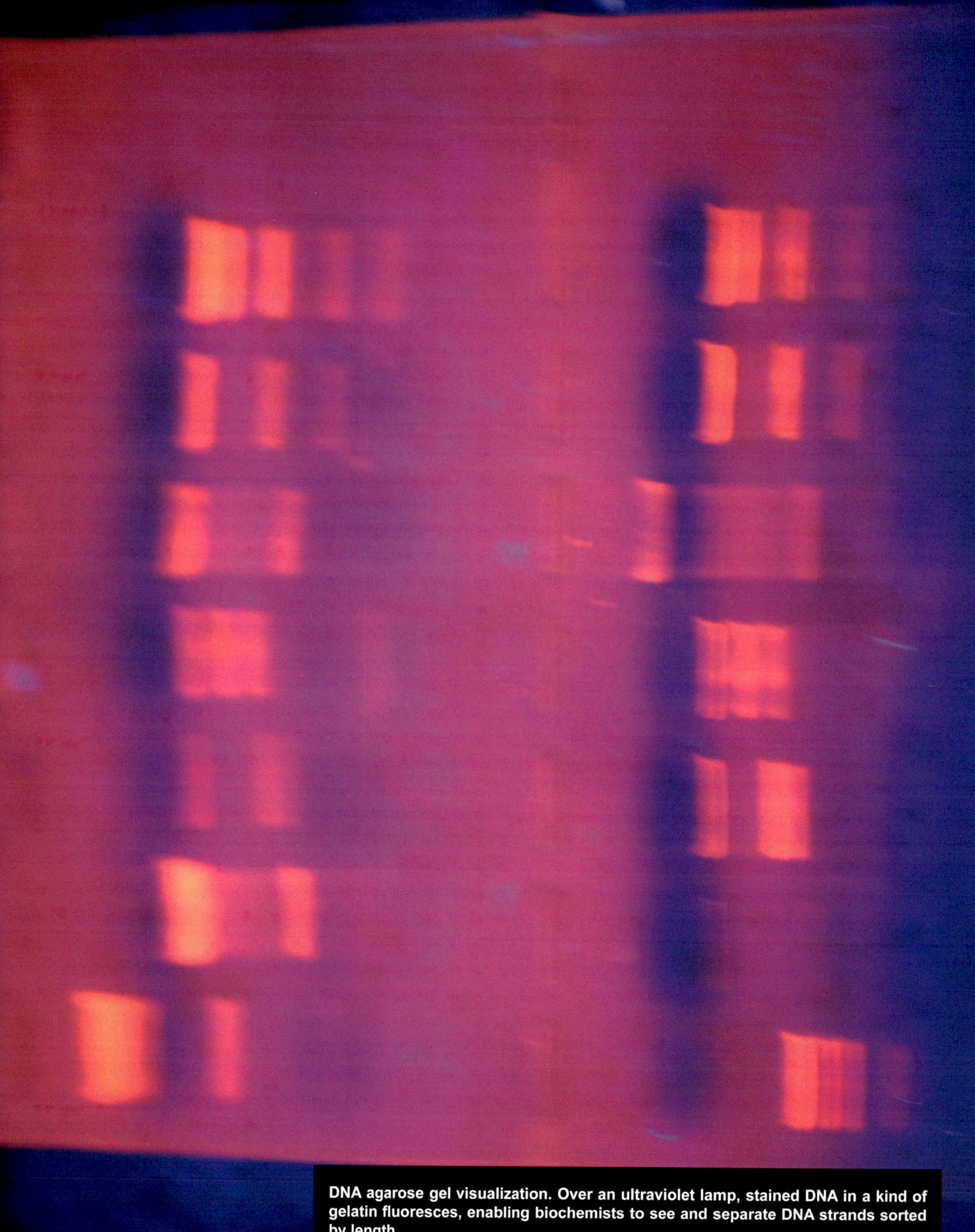

DNA agarose gel visualization. Over an ultraviolet lamp, stained DNA in a kind of gelatin fluoresces, enabling biochemists to see and separate DNA strands sorted by length.

Chapter 9

The Biology of Stem Cells

Nathaniel T. Jeanson, Ph.D.

"Embryonic stem cell research kills babies!" "Stem cell research could save your life!" We are all familiar with provocative claims such as these associated with stem cell research. But what are stem cells and why are they so controversial?

Previous chapters presented two cellular distinctions based on subcellular architecture (prokaryotic versus eukaryotic, and plant versus animal). The term "stem cells" implies a third cellular distinction—one based on cell *potential*. Stem cells are distinct from non-stem cells in their potential for producing multiple cell types. The fact that multiple cell types exist in the human body implies a fourth cellular distinction—one based on cell *function*. These latter distinctions are crucial for understanding why stem cells exist and why they may be useful therapeutically.

The Biology of Adult Stem Cells

The purpose and function of *adult stem cells* (aSCs) are best understood in the context of adult tissue and organ physiology, and also in the context of how the non-stem cells of the adult combine to sustain this physiology.

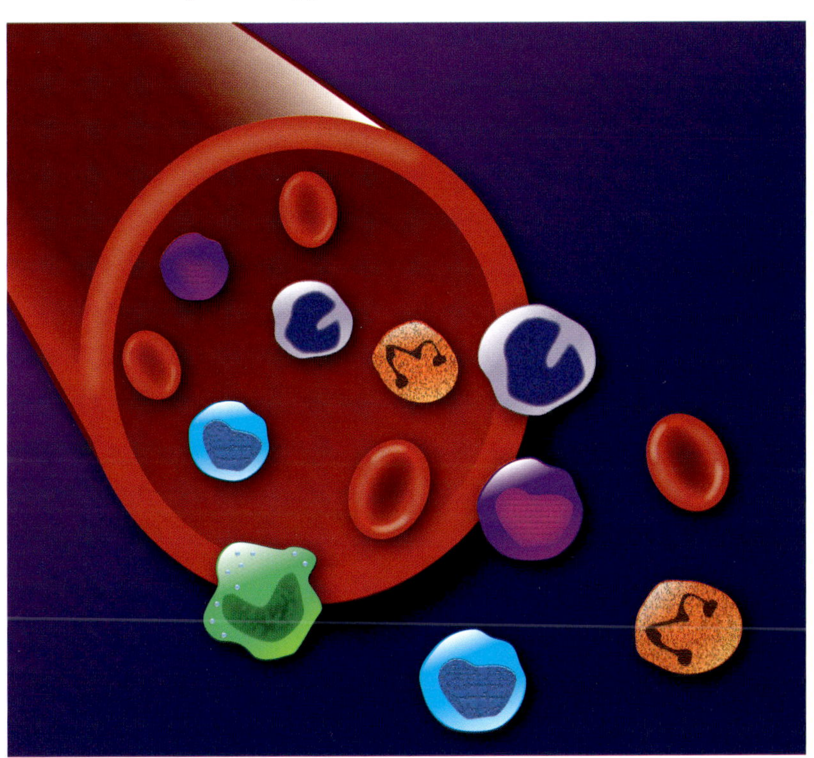

The human body is made up of trillions of eukaryotic cells. However, not all of these cells are the same. Profound differences are readily apparent when various cells are observed under a microscope. Smooth muscle cells of the heart are clearly distinct from the rod and cone cells of the eye. Fibroblasts of the connective tissue are noticeably different from the sperm and egg cells of the reproductive system. Astrocytes and neurons in the brain are easily distinguishable from the red and white cells of the blood system. These visual differences reveal the enormous cellular diversity that exists just within the human body. There are also many differences in cells between different types of animals.

The different cell types of the human body

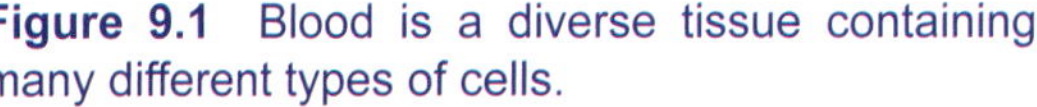

Figure 9.1 Blood is a diverse tissue containing many different types of cells.

combine to form the tissues and organs that enable a person to breath, move, and function. Individual organs of the digestive system—such as the small intestine, large intestine, and pancreas—consist of many cell types that work in concert to convert the chemical energy in food to energy that the body can use to do work. The transport system for cells and energy is blood, which is itself a diverse tissue, consisting of red blood cells, several types of white blood cells, and cells that participate in the clotting cascade (Figure 9.1). Together, this functional diversity makes the actions of the human body possible.

In a healthy human adult, different cell types have different lifespans. For example, red blood cells live for three months, while some white blood cells may survive for only eight hours. Thus, the bloodstream is constantly changing as blood cells are continuously dying or being removed. Compounding the issue of tissue turnover in healthy individuals is the cell death that results from damage. Bacterial and viral infections, chemical pollutants, and physical injury all acutely decrease the number of cells present in the body. These stresses on the body can affect nearly any organ, not just blood. Without a mechanism to replace lost and dying cells, the body would soon shrivel and collapse due to excessive cell loss.

Most cells of a tissue or organ cannot or do not sustain tissue turnover. Many cell types, such as muscle fibers and neurons, do not divide. Red blood cells do not even possess a cell nucleus, making it impossible for them to divide. And of the cell types that engage in division, many do not possess the potential to form other cell types. For example, some types of white blood cells can be stimulated to divide by bacterial infection; however, these white blood cells produce only white blood cells—they do not produce red blood cells. The limited ability of most adult cells to divide or produce cellular diversity severely limits the capacity of an organ to sustain itself.

Theoretically, a specialized cell type dedicated to cell division and to production of cellular diversity would solve the problem of tissue renewal. Furthermore, for the tissue to be maintained indefinitely, this specialized cell would require the ability to divide indefinitely—dividing to make more of itself, as well as more of the depleted cell type. Stem cells are the specialized cell type that solves the problem of tissue renewal.

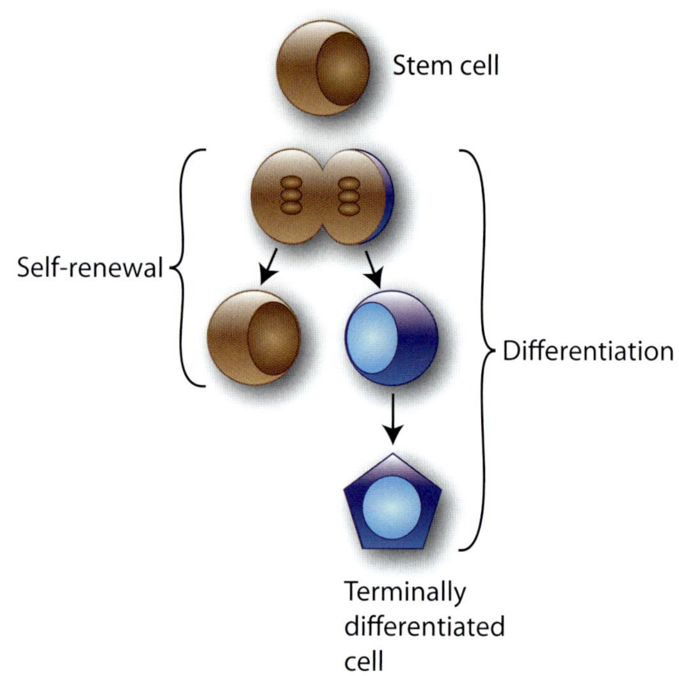

Figure 9.2 Diagram illustrating the processes of stem cell self-renewal and differentiation as a way to sustain tissue.

Stem cells are defined by two properties: *self-renewal* and *differentiation* (Figure 9.2). Self-renewal is the celll's ability to make more of itself. Differentiation is the process by which one cell type (defined by function—for example, a blood progenitor cell) becomes another cell type (also defined by function—for example, a red blood cell).

Many cells lack both of these properties. Other cell types possess only one of these qualities. For example, some precursor cells of the blood system (i.e., cells derived from stem cells but not yet fully differentiated) can differentiate into multiple cell types of the blood system but possess little, if any, capacity for self-renewal. Only stem cells, by definition, possess these two characteristics.

Stem cells are dedicated solely to cell division and to the production of the other cell types that do the "work" of the tissue. For example, blood stem cells do not transport oxygen or fight infection, but they do participate indirectly in these processes by differentiating into red and white blood cells that perform these functions. For this reason, stem cells are referred to as "primitive" or "undifferentiated" to signify their lack of commitment to a particular task. Stem cell progeny are referred to as "committed" or "mature" to signify their direct participation in the work of the tissue. These characteristics make stem cells unique in their function and purpose.

Stem cells exist in many, but not all, tissues of the human body. As a general rule, stem cells are found in tissues with high cellular turnover, such as bone marrow, skin, gut, or testes. In tissues with low or no cellular turnover, such as the liver, brain, and pancreas, the existence of stem cells is contested. Thus, the human body's ability to repair itself seems to be limited to a subset of highly proliferative tissues.

Identifying the precise tissue distribution of stem cells in the adult is difficult due to the unique biological properties of stem cells. First, in tissues where stem cells have been identified, stem cells exist in low numbers. The majority of cells in these tissues consist of mature cells, not stem cells. Thus, attempting to identify stem cells in a tissue is an involved process. Second, since stem cells exist to divide rather than to perform a particular function, they lack obvious identifying characteristics, such as the red color of red blood cells. Therefore, isolation of stem cells has been limited to techniques involving either cell surface markers and proteins or the anatomical position of a cell in a tissue.

Furthermore, since stem cells are defined by their cell division potential, stem cell identification ultimately requires cell transplantation. All of these experimental tests are slow and labor-intensive, making stem cell identification and isolation very time-consuming and challenging. Hence, it is theoretically possible that stem cells exist in all tissues but that current technological limitations prevent their identification.

In the adult, stem cells produce mature cell types specific to the tissue from which they originate. While the progeny of stem cells may migrate to other tissues, the cells themselves do not morph into other cell types. For example, blood stem cells differentiate into only blood cells; they do not differentiate into gut or skin cells. Likewise, skin stem cells form only skin, not brain or another tissue. While initial experiments suggested that stem cells were much more plastic and possessed a potential to form mature cells for a variety of tissues, these findings have been difficult to replicate. Thus, aSCs sustain the cellular turnover of only the tissue in which they reside.

In summary, the human body possesses incredible cellular and tissue diversity that is sustained by stem cells, tissue-specific cell types characterized by the unique properties of self-renewal and differentiation.

Cellular Mechanisms of Self-Renewal and Differentiation

A fuller understanding of aSCs, their function, and their potential for therapeutic use requires an understanding of the cellular and molecular mechanisms involved in self-renewal and differentiation. The outcome of any cell division produces two cells. For stem cells, there are two theoretical outcomes of cell division (Figure 9.3). First, a stem cell can divide into two identical daughter cells, either two stem cells or two mature cells. This is referred to as *symmetric* cell division, indicating that both daughter cells have the *same* fate. Sec-

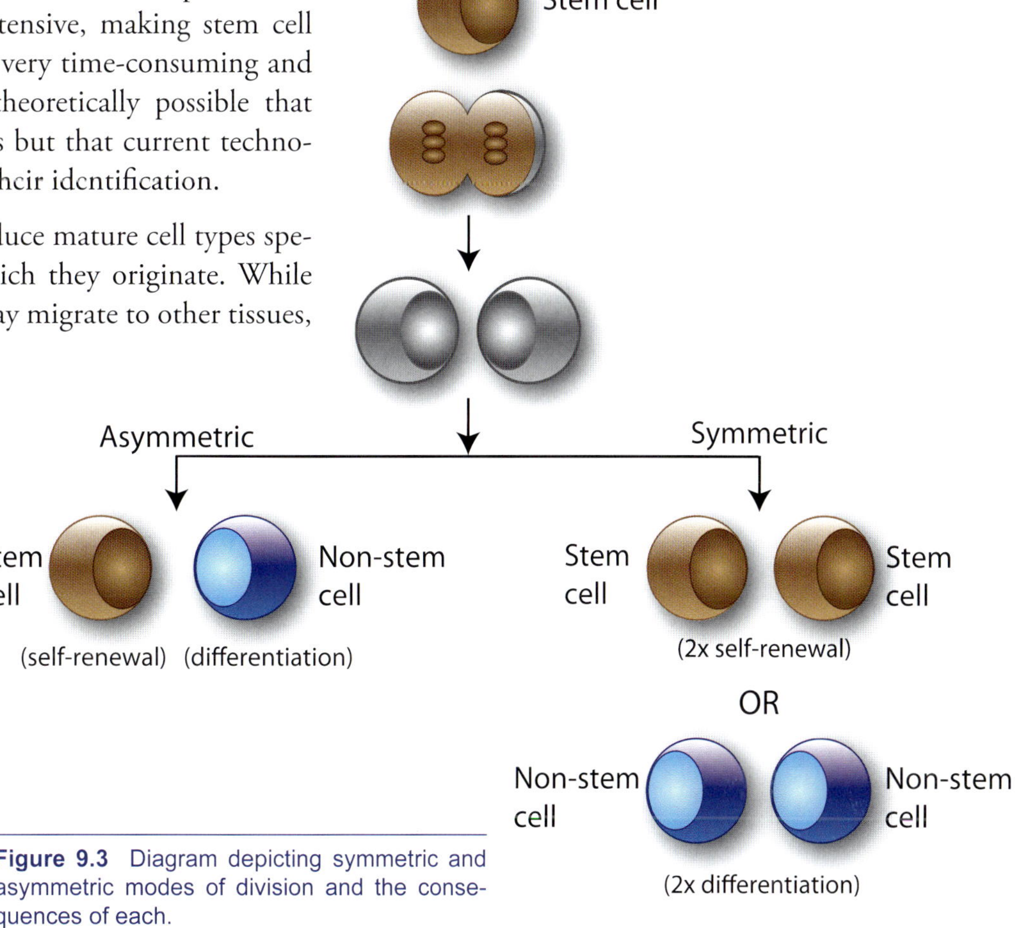

Figure 9.3 Diagram depicting symmetric and asymmetric modes of division and the consequences of each.

ond, stem cells can undergo *asymmetric* cell division, producing two daughter cells that have *different* fates (one stem cell and one mature cell). This distinction in cell division applies to stem cells in all tissues.

The appropriate combination of symmetric and asymmetric stem cell division sustains tissue turnover. A pattern of purely asymmetric stem cell division would theoretically maintain the status quo in a tissue, since stem cell numbers would be unchanged and mature cells would be continuously produced. However, in situations of injury when stem cell numbers are decreased, asymmetric stem cell division would not lead to recovery of the original stem cell numbers. In these circumstances, asymmetric stem cell division must be combined with symmetric self-renewing division (production of two daughter stem cells) in order to recapture the original number of stem cells in the tissue.

Conversely, when mature cell numbers are severely depleted, it becomes necessary for stem cells to produce mature cells at an increased rate through symmetric differentiative division that results in two mature cells. However, this latter mode of division cannot sustain a tissue long-term, as it depletes stem cell numbers. Thus, to sustain tissue turnover in health and injury, stem cells must carefully balance self-renewal and differentiation by an appropriate combination of symmetric and asymmetric cell division.

Molecular Mechanisms of Self-Renewal and Differentiation

The control of symmetric and asymmetric cell division occurs at the molecular level. Precise gene regulation (various genes either turned on or off) directs adult stem cell self-renewal and differentiation outcomes. While all cell types—primitive and mature—possess the same set of genes, these genes are used very differently depending on the function of the cell in question. Some of the genes used by a particular cell type are known.

For example, turning on the *gata-1* gene in blood progenitor cells leads them toward a red blood cell. In its genome, stem cells possess the same genes found in a mature cell. Therefore, for a stem cell to maintain its present identity, it must *suppress* the activation (turning on) of genes (such as *gata-1*) that promote mature cell development. However, the genes involved in the regulation processes (turning genes on and off) at the stem cell stage are largely unknown. Nevertheless, there exists a means by which stem cell fate is regulated intracellularly.

The intracellular programs that control the fate of stem cells depend heavily on signals received from their environment, such as various cell types within the tissue and within the body. These signals direct stem cell differentiation in order to maintain the tissue in which they reside when mature cell numbers are high or low. A number of these differentiation-inducing signals have been identified.

For example, the hormone erythropoietin acts at the surface of primitive cells to stimulate an intracellular program that leads to red blood cell production. Self-renewal-inducing signals are necessary in situations when stem cell numbers are low. However, the identity of these signals remains an outstanding problem. Although the molecular mechanisms are not fully understood, stem cells interact with and respond to their environments via very specific molecular signals.

The exquisite control that the body exercises over stem cell self-renewal and differentiation speaks to the efficient and intelligent design of tissue maintenance processes. Despite our ignorance of all the molecules that determine stem cell fate, the body sustains tissue turnover without interruption in billions of individuals every day. The magnitude of this biological requirement is astonishing.

Consider just red blood cells. Because red blood cells have only a three-month lifespan, in order to operate effectively the blood system must replace millions of red blood cells each day to maintain proper red blood cell counts. One malfunction in this system and the tissues of the body would quickly suffocate due to lack of oxygen. Yet death due to malfunction in red blood cell production is rare. If the body is sufficiently robust and efficient to sustain this essential system so consistently, the molecular controls of this process must be very complex. Furthermore, if hundreds of intelligent researchers are still stymied in their ability to elucidate this logic—let alone design it on their own logic—how much more intelligent must the Creator of this system be!

In summary, adult stem cells perform tissue maintenance by the processes of self-renewal and differentiation, which are executed by largely unknown mechanisms in a very robust and consistent manner. Time and chance are certainly not the creators. They are far *inferior* to our limited human intelligence, and vastly more inferior to the intelligence required to *design* sustainable systems like the tissues of the body.

The Biology of Embryonic Stem Cells

The function and purpose of adult stem cells is very different from that of embryonic stem cells. As the discussion above demonstrates, the cellular diversity of the human body is enormous. The existence of this complexity immediately raises an important question: What is its origin? The answer to this question is vital to understanding the purpose and function of embryonic stem cells. As the name implies, *embryonic stem cells* (ESCs) are derived from early human embryos. Hence, to understand ESC biology, it is necessary to first discuss early human development.

Human life begins when a single sperm cell from the father unites with a single egg from the mother to form a *zygote*. The zygote develops into a fetus, and, after nine months, a newborn is delivered (Figure 9.4). The progression from zygote to fetus to newborn occurs through two fundamental processes: cell division and cell differentiation. Clearly, for a single cell to form a baby and then grow into an adult requires billions of cell divisions. Furthermore, a very tightly controlled process of differentiation is required for a single cell to become hundreds of different cell types. These two operations define embryonic development, fully realized in the generation of the adult organs and tissues. Thus, the intricacy of the human body is rebuilt from a single cell every time a human being is conceived.

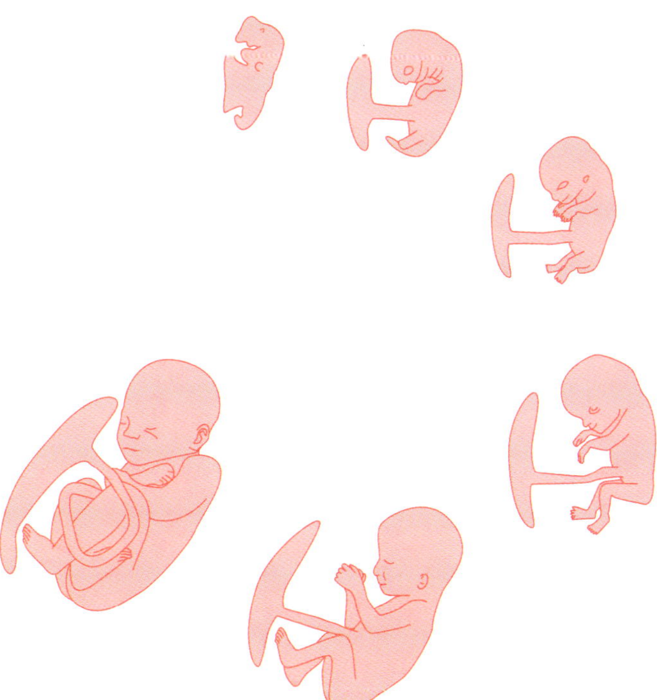

Figure 9.4 Diagram depicting various stages of human embryonic development.

The molecular control of this developmental process is poorly understood. If aSC differentiation into its mature progeny is complicated, then the millions of steps from a zygote to a newborn and then to an adult must be nearly inconceivable. Enormous amounts of research will be required to unravel how the developmental program hard-wired into the DNA molecule (the molecule of heredity) transforms a single cell into the complexity that is an adult human.

How Embryonic Stem Cells Work: Mechanisms of ESC Self-Renewal and Differentiation

ESCs possess the two defining attributes of stem cells: self-renewal and differentiation. A full understanding of ESC potential and use requires a discussion of the molecular mechanisms involved in ESC self-renewal and differentiation.

The cellular mechanisms of ESC self-renewal and differentiation are unique in that ESCs do not divide asymmetrically. All ESC cell divisions appear to be symmetric, resulting in either two daughter stem cells (a self-renewal division) or two daughter differentiated cells (a differentiative division). Thus, to maintain a population of ESCs in the culture dish, there must be sufficient numbers of symmetric self-renewal divisions.

The molecular signals that control ESC differentiation are poorly understood. ESCs retain an immense capacity for differentiation into diverse cell types. This powerful potential is also the reason that the mechanisms of ESC differentiation are difficult to decipher. With so many potential fates, the number of molecules that could influence ESC differentiation is likely enormous. Thus, like adult stem cells, ESC differentiation remains an outstanding biological problem.

Currently, there is a lack of knowledge regarding the molecules that govern aSC self-renewal. In contrast to this ignorance, the molecular signals that control ESC self-renewal are better understood. This understanding likely stems from several differences between ESCs and aSCs.

First, the *inner cell mass* (ICM) from which ESCs are derived (in a process explained more fully in the next chapter) has a different biological purpose than aSCs; thus, the signaling mechanisms that control ESC self-renewal would be predicted to be different from and independent of aSC self-renewal signals.

Second, unlike aSCs, ESCs are easily identifiable and obtainable. Since the *blastocyst* (the hollow ball stage of early embryonic development) is small and there are few steps from the zygote to the blastocyst stage, it is relatively easy to see the ICM microscopically and obtain ICM cells for ESC culture. Hence, a readily available cell source enables the research needed to identify ESC self-renewal signals to proceed more rapidly.

Third, the ICM from which ESCs are derived represents a high percentage of the blastocyst (source tissue). This is in direct contrast to the rare frequency of stem cells in the adult, making aSC isolation and study difficult and ESC isolation relatively easy. Together, these properties of ESCs have advanced our understanding of ESC self-renewal far beyond our knowledge of aSC self-renewal processes. Indeed, ESCs are readily propagated in tissue culture, a phenomenon unrealized for aSCs.

Thus, ESCs are a unique population of stem cells that are derived from the ICM of the blastocyst. They retain the full developmental potential of the ICM. This developmental power is exhibited when ESCs are placed in the appropriate cellular or chemical contexts.

In summary, though ESCs and aSCs are both defined by the ability to self-renew and differentiate, it should be apparent that these two types of stem cells are fundamentally different. First, they differ in their source. ESCs are derived from the ICM of early development, while aSCs are derived from adult tissues. Second, these stem cells differ in their differentiation potential. Adult stem cells exist to maintain tissue turnover in the tissue in which they reside, limiting their differentiation potential to cells of their tissue of residence. In contrast, the ICM from which ESCs are derived exists to generate every type of tissue in the adult; hence, ESCs theoretically have unlimited differentiation potential. Third, ESCs and aSCs are drastically different in their in vitro (lab-based) propagation potential. ESCs can be cultured in the dish virtually indefinitely; aSCs cannot. These profound differences clearly distinguish these cell types, but they do not prevent their interconversion (i.e., the conversion of one into the other).

The Origins of Human Cellular Diversity: The Fact of Creation and the Insufficiency of Evolution

The first human zygote could not have evolved from an earlier life form via the process of evolution. This statement holds true on several levels of evolutionary thought. First, a human zygote could not have resulted from modification of a chimpanzee-like or primate-like zygote. Under the evolutionary paradigm, chimpanzees are our closest animal relatives, and the supposed 98 to 99 percent genetic identity between humans and chimpanzees is touted as evidence of this relationship. Even if we grant the evolutionists that percentage of similarity, this argument is quickly turned on its head when we consider the vast behavioral differences between us and these apes. The atheistic evolutionist has the unenviable task of explaining all these behavioral differences *with a very small fraction of genetic differences*—a seemingly insurmountable problem.

Second, the human zygote could not have evolved from an "earlier" non-primate animal, such as a reptile-like or fish-like creature. This is evident when developmental biology is examined in light of Scripture. Genesis 1 clearly teaches that God created distinct "kinds" of creatures, and Genesis 6 and 7 imply that there exists a *limit* to biological change.

For instance, Noah took two of *every* kind of land-dwelling, air-breathing creature to preserve them post-Flood, not two of *some* kinds, implying that kinds cannot be changed into other kinds. Creatures (and, likely, biblical kinds) can be distinguished based on morphology (the form and structure of an organism), and morphology is rebuilt during development every generation. Therefore, the limit to biological change among kinds is likely found in the developmental programs of each creature. Indeed, fish embryonic development can be distinguished from mammalian embryonic development at the four-cell stage—just two cell divisions after the meeting of sperm and egg. While the limit to biological change has not yet been formally identified, these preliminary data are consistent with the fact that the human zygote could not have evolved step-by-step from a fish-like zygote.

Third, the human zygote could not have evolved from an "earlier" asexually reproducing creature. To transition from an asexual mode of reproduction to a sexual mode requires the addition of new parts to the system—separate male and female cells. Evolutionary theory postulates that increasing complexity arises as a result of natural selection favoring certain mutations over others. However, how could natural selection favor mutations that gave rise to an "egg" when no sperm had yet evolved to fertilize it? This egg would be useless without a sperm—a seemingly highly *unfavorable* condition that natural selection should eliminate. For sexual reproduction to have evolved, both sperm and egg must have "miraculously" appeared together at the same time. Aside from an evolutionary "miracle," the human zygote could not have arisen by the process of evolution.

Thus, the variety of cells that work together to sustain the human adult originates from a single cell whose ultimate origin is God, not evolutionary descent with modification from a "simpler" reproductive program.

Wisdom, forethought, and intelligent engineering are the obvious logical conclusions from the observation of human developmental efficiency and consistency. The more than seven billion people alive today and the vast numbers who lived in the past testify to the robustness and superior design of the plan that must exist in the developmental logic. Furthermore, the inability of hundreds of intelligent scientists to discover the molecular logic of development (let alone design their own) betrays the existence of an extremely intelligent Creator responsible for the human reproductive program.

Summary

1. *Stem cells* have the potential to produce other cell types. Many of the cells in tissues and organs cannot renew themselves. Stem cells are able to make both more of themselves (*self-renewal*) and other cell types (*differentiation*), solving the problem of tissue renewal.
2. *Adult stem cells* produce mature cells that are specific to a particular tissue type—blood stem cells produce blood cells, skin stem cells produce skin cells, etc. Stem cells must carefully balance how they divide in order to maintain healthy levels of stem cells and mature cells.
3. *Embryonic stem cells* are different from adult stem cells in that they can form every cell type of the body. Thus, they can generate the many kinds of tissues needed to grow a *zygote* (the single cell from which a baby develops) into a full-term baby and then to an adult. The molecular control needed to do this must be almost unbelievably specific and complex.
4. Embryonic stem cells are better-understood than adult stem cells because they are easy to identify and obtain, and can readily be grown in the lab. Adult stem cells are not only relatively scarce, they can be difficult to isolate. They also lack the developmental potential of embryonic stem cells.
5. The first human zygote could not possibly have evolved from a chimpanzee-like or earlier non-primate animal. The many types of cells that work together to sustain the human adult are developed from a single cell whose ultimate origin is God. Only an extremely intelligent Creator can explain the existence of this marvelous system.

Darwinian Medicine: A Prescription for Failure

Randy J. Guliuzza, P.E., M.D.

Although the field of medicine experienced major advances in the 20th century, it also embraced mistakes and outright atrocities caused by the introduction of evolutionary thinking. Setbacks to the profession are bad, but thousands of patients "treated" with Darwinian medical principles suffered needlessly, experiencing confusion, painful surgery, and even death.

Darwinism Promoted the Medical Practice of Eugenics

In the most egregious example, many physicians advocated the promotion and practice of eugenics. In a quest to improve the overall genetic composition of the human race, eugenicists selectively bred biologically "superior" people and forcibly eliminated genetic defects by sterilizing, aborting, or euthanizing "inferior" people.

This practice can be laid squarely at the feet of Darwinian medicine.[1] Many lives were destroyed through the first large-scale manifestation of Darwin's belief that "the civilized races of man will almost certainly exterminate, and replace, the savage races throughout the world."[2]

Darwinism Advocated Needless Surgical Procedures to Remove "Vestigial Organs"

Even patients fortunate enough to be deemed "fit" still might not avoid the surgical knife. Because of Darwin's *The Descent of Man*, the appendix became widely regarded as a worthless rudimentary organ left over from man's herbivorous ancestors. This led to a decades-long fundamental flaw in Darwinian medicine: the expectation that people would be better served without certain organs, even perfectly healthy ones.

By the mid-20th century, thousands of "prophylactic" surgeries had been performed based on assumptions such as "the sooner [vestigial appendages] are removed the better for the individual."[3] A 2007 Duke University Medical School press release challenged this naïve view: "Long denigrated as vestigial or useless, the appendix now appears to have a reason to be—as a 'safe house' for the beneficial bacteria living in the human gut."[4]

A biochemistry professor stated that this possible bacterial function "makes evolutionary sense."[5] The medical fate of the appendix remains uncertain, but fortunately, published medical advances now advocate that the removal of tonsils be contingent on meeting evidence-based medical criteria[6]—a feature totally lacking in Darwinian medicine.

Darwinian Prejudices Have Hindered Medical Research

Darwinian medicine's concept of vestigial organs has also retarded medical research, since there is little incentive to study "useless" structures. This mistaken belief has permeated even the cellular and molecular levels. Stanford University reported in 1998 on certain white blood cells that heretofore had been largely ignored by immunologists. Why? The "natural killer" (NK) cells were "thought by some to be an archaic remnant of the primitive mammalian immune system."[7] Other areas of profitable medical research continue to be held back by the smothering assumptions of Darwinian medicine.

Darwinian Medicine Courses Lack Medical Significance

Ignoring Darwinism's bad medical track record, some scientists stridently advocate the introduction of a new course in medical school: Darwinian Medicine. Two vocal proponents assert:

> Evolutionary biology...has not been emphasized in medical curricula. This is unfortunate, because new applications of evolutionary principles to medical problems show that advances would be even more rapid if medical professionals were as attuned to Darwin as they have been to Pasteur.[8]

Pasteur's contributions to medicine, which were completely independent of evolutionary assumptions, are legendary. He directed research into areas that have undeniably saved millions of lives. Compared to Pasteur's seminal research, the Darwinian approach to medicine and its derived explanations are insignificant. Evolutionary biologist Paul Sherman advocates an approach that examines whether symptoms are "useful adaptations" or true pathologies.

> For example, a mild fever...is often the body's natural response to infection. Studies show that a mild fever leads to faster recovery

times....[Sherman] noted that a Darwinian medicine approach adds to the doctor's toolbox to offer a wider range of treatments, including advising a patient in some instances to help the body's evolved system do the healing.[9]

Cutting-edge Darwinian theories on illness include: 1) X-linked color blindness evolved to help male paleolithic hunters see camouflage; 2) the itch associated with insect bites evolved so people would avoid being bitten; 3) myopia may result from an interaction between genes and the close work characteristic of literate societies; 4) salivation, tearing, coughing, sneezing, vomiting, and diarrhea evolved to expel noxious substances and microbiologic agents; and 5) humans' natural repugnance toward garbage, feces, vomitus, and purulence is an evolved defense against contagion.[10]

But the Darwinian method amounts to little more than making observations of signs and symptoms and appending unsubstantiated evolutionary stories as window dressing. Trite explanations are published in collaborative peer-reviewed journals in articles that contain only a tiny fraction of the scientific rigor of medical articles featured in the *Journal of the American Medical Association*. These "insights" are hardly on a par with the significance of Pasteur's work.

Darwinian Medicine Has No Clinical Value

Darwinian medicine adds nothing to the doctor's toolbox. For instance, the only Darwinian aspect to Sherman's interpretation of the observed infection-fever interaction is his fully inexplicable assumption that fever is an evolved response. Most physicians already knew what fevers to treat.

Such explanations fail accepted scientific standards since they cannot be tested. Moreover, serious medical researchers would not invest time in them since their medical contribution is negligible. Even putative beneficial observations of natural selection such as bacterial resistance to antibiotics and the heterozygotic advantage of sickle cell disease are not based on Darwinian medicine, but were observed through the relevant basic sciences of microbiology and molecular genetics.

Darwinian Medicine Has No Predictive Value

None of the Darwinian explanations integrate (much less are based on tests of) the phylogeny or actual physical evolutionary development of the organism itself. Given the long time to develop new drugs, a real test would be a Darwinist prediction—based solely on human evolutionary phylogeny—of a new, presently unobserved disease for which pharmaceutical companies should start developing a treatment. So far, no such predictions have been forthcoming.

This failure, coupled with increased needs to teach new medical research, is possibly why evolutionary medicine is currently squeezed out of every American medical school's curricula. "Add to this the fact that the field has failed so far to provide clinically useful findings and you see why medical schools lack interest," admitted evolutionary medicine proponent Stephen Lewis.[11]

Conclusion

Many of the giants in medicine—Edward Jenner, Gregor Mendel, Louis Pasteur, Howard Florey and Ernst Chain, Selman Waksman—did pioneering work while either rejecting Darwinism or ignoring it altogether. Darwinian medicine is a sham. It stands on the backs of real researchers, dresses up their major medical insights with evolutionary stories, and then claims them as its own, while diverting grant money away from bona fide medical research.

The legacy of Darwin's ideas to medicine ranges from irrelevant to disastrous. But beyond the wasted time, talent, and resources of the medical community, Darwin's most lasting legacy to the field may well be the suffering of those whom medicine was originally meant to heal.

References

1. Weikart, R. 2004. *From Darwin to Hitler: Evolutionary Ethics, Eugenics, and Racism in Germany.* New York: Palgrave Macmillan. See also *War Against the Weak: Eugenics and America's Campaign to Create a Master Race.*
2. Darwin, C. 1901. *The Descent of Man.* London: John Murray, 241-242.
3. Rabkin, W. The Pros and Cons of Tonsillectomy. *South African Medical Journal.* 8 January, 1955, 30.
4. Appendix Isn't Useless at All: It's a Safe House for Bacteria. Duke Medicine press release, October 8, 2007.
5. Appendix May Produce Good Bacteria, Researchers Think. Associated Press, October 5, 2007.
6. Clinical Indicators Compendium. American Academy of Otolaryngology-Head and Neck Surgery.
7. Weidenbach, K. Natural-born killers: An immunologic enigma solved. *Stanford Report.* Stanford University news release, January 14, 1998.
8. Williams, G. and R. Nesse. 1991. The Dawn of Darwinian Medicine. *The Quarterly Review of Biology.* 66 (1): 2.
9. Ramanujan, K. Intelligent design? No smart engineer designed our bodies, Sherman tells premeds in class on Darwinian medicine. Cornell University press release, December 7, 2005.
10. Rannala, B. 2003. Evolving Health: The Origins of Illness and How the Modern World Is Making Us Sick. *Journal of the American Medical Association.* 289 (11): 1442-1443; Nesse, R. M. and G. C. Williams. 1994. *Why We Get Sick.* New York: Random House.
11. Baker, M. Darwin in medical school. *Stanford Medicine Magazine.* Summer 2006.

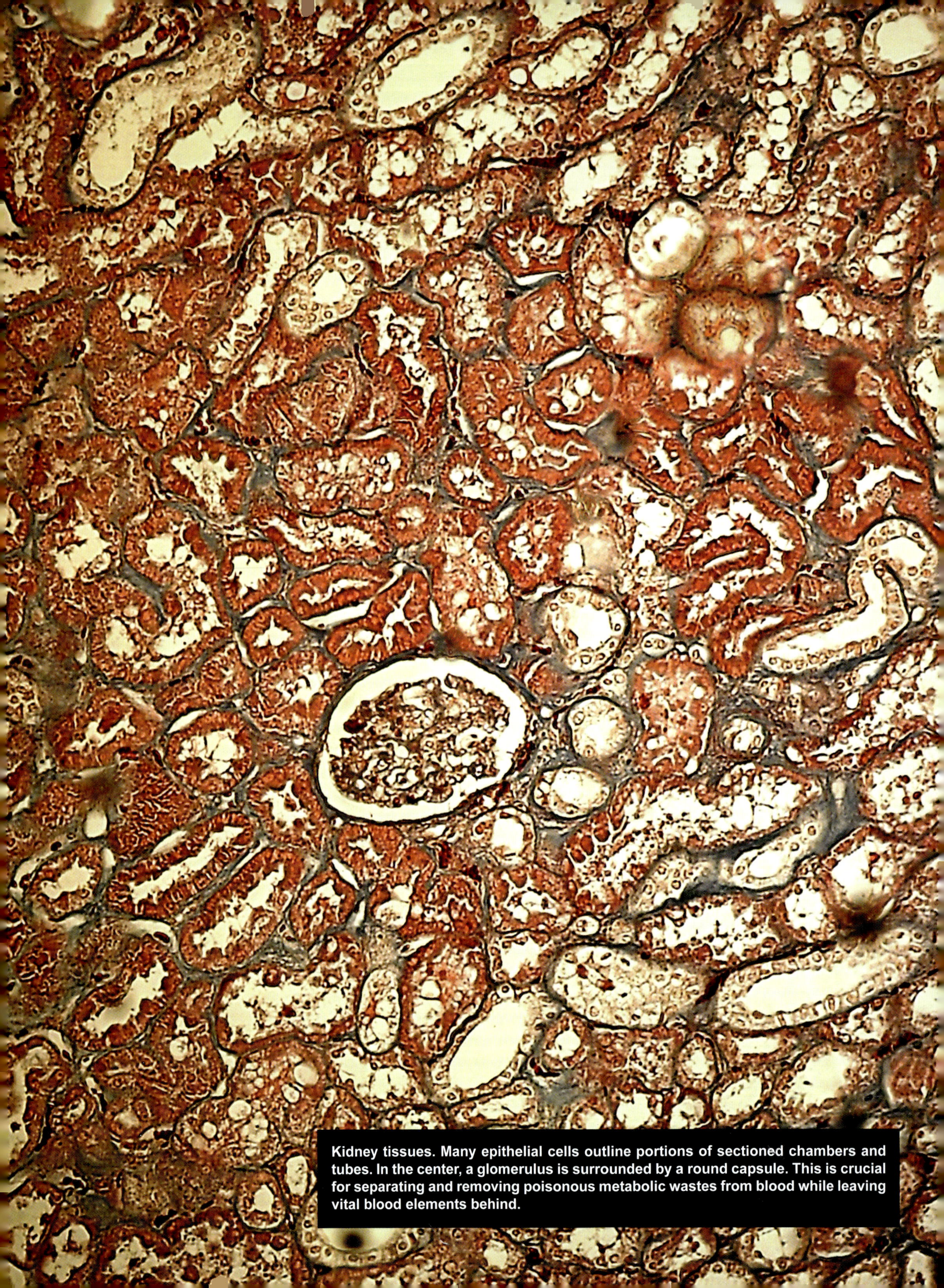

Kidney tissues. Many epithelial cells outline portions of sectioned chambers and tubes. In the center, a glomerulus is surrounded by a round capsule. This is crucial for separating and removing poisonous metabolic wastes from blood while leaving vital blood elements behind.

Chapter 10

Processes and Implications of Stem Cell Research

Nathaniel T. Jeanson, Ph.D.

Embryonic stem cells (ESCs) are derived from the earliest stages of the human developmental process. These early stages are readily identifiable. After the fusion of sperm and egg, the newly formed *zygote* cleaves once to form the two-cell stage. Another cleavage of each daughter cell follows, forming the four-cell stage. Division of all four daughter cells then gives rise to the eight-cell stage. The morula and blastocyst stages result from further division. It is the *blastocyst* stage from which ESCs are derived.

The blastocyst stage has two distinct cell populations. The outer wall of cells, termed the *trophoblast* layer, is responsible for implantation of the blastocyst into the uterus, and it eventually gives rise to the placenta. On the inner side of the trophoblast layer is a group of cells, the *inner cell mass* (ICM), which eventually forms the fetus. Thus, the ICM and the trophoblast layer are temporary stages that are unique in their developmental potential. Neither cell population persists in the fetus or into adulthood. Rather, each exists to spawn further cell populations that eventually differentiate into the cells that form the diverse tissues of the baby and adult.

To harvest ESCs, the blastocyst is destroyed and the ICM is removed (Figure 10.1). Once harvested, the ICM is grown on a cell layer or in a defined chemical mix that causes the cells to continue cell division but stops cell differentiation. The resultant cell line is termed an *ESC line*. Thus, the ICM, not the trophoblast layer, spawns the ESCs.

While ESCs are *derived* from the ICM, these two cell types may, in fact, be identical. In the mouse, it is possible to create a blastocyst-stage embryo without a functional ICM. ESCs can be injected into an ICM-less blastocyst and the resultant embryo implanted into a surrogate mother. This manipulated embryo will

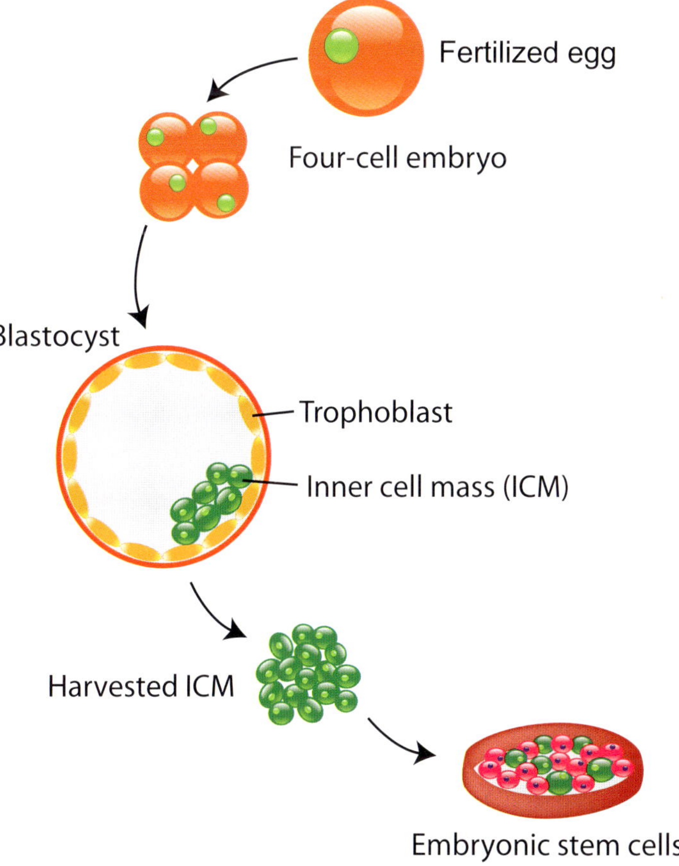

Figure 10.1 Embryonic stem cells are harvested by destroying the blastocyst and removing the inner cell mass.

develop into an adult that is derived *entirely* from the donated ESCs. Thus, when re-exposed to the appropriate environment, a cluster of ESCs will behave like an ICM, suggesting that these two cell populations are equivalent.

While the process of ESC derivation destroys the blastocyst, the ESC state appears to be a *suspension* of development rather than a *termination* of development. Normally, during blastocyst development the ICM differentiates on command and ultimately forms the

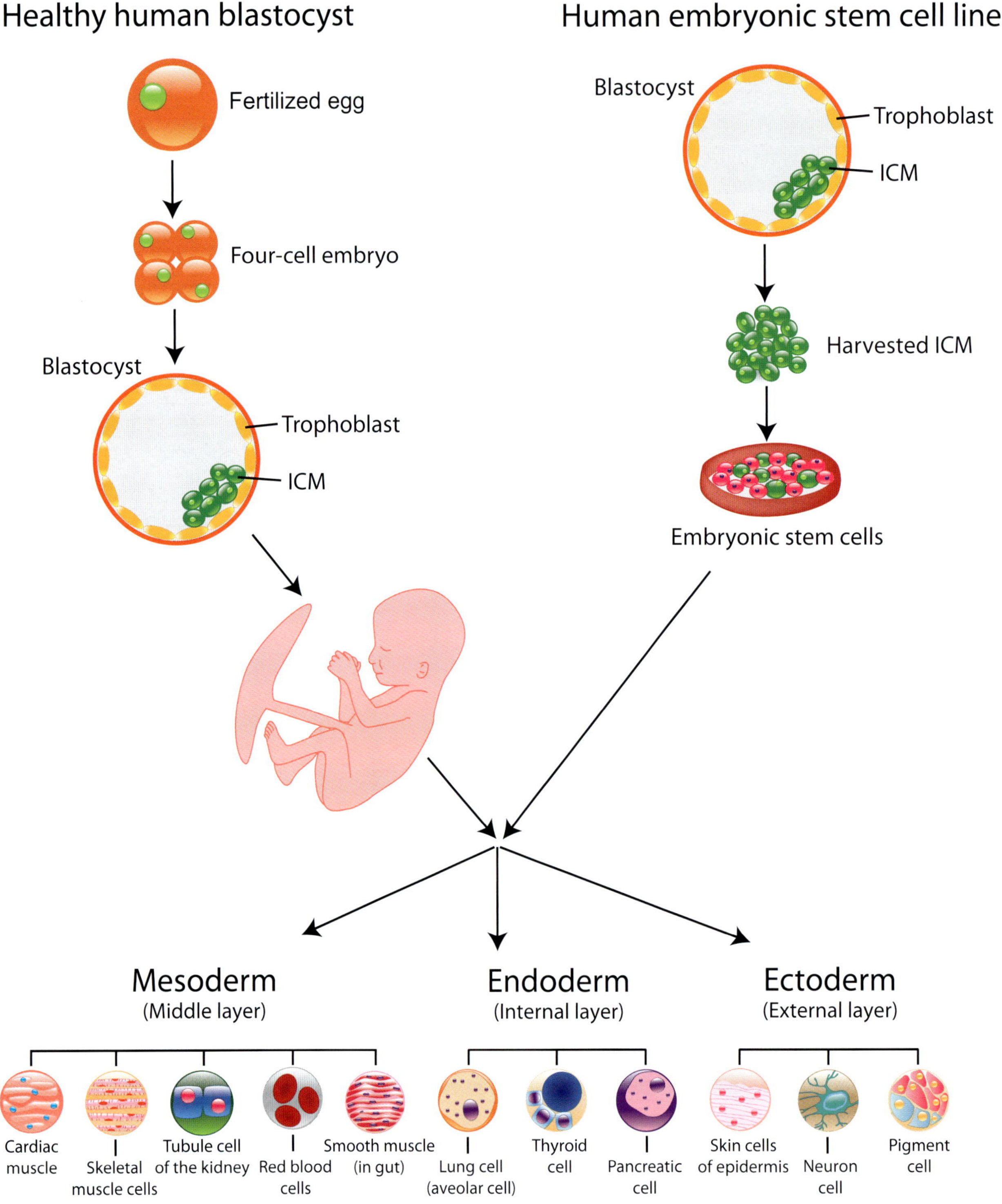

Figure 10.2 Embryonic stem cells have the potential to differentiate into all three groups of cells that form during the development of an embryo.

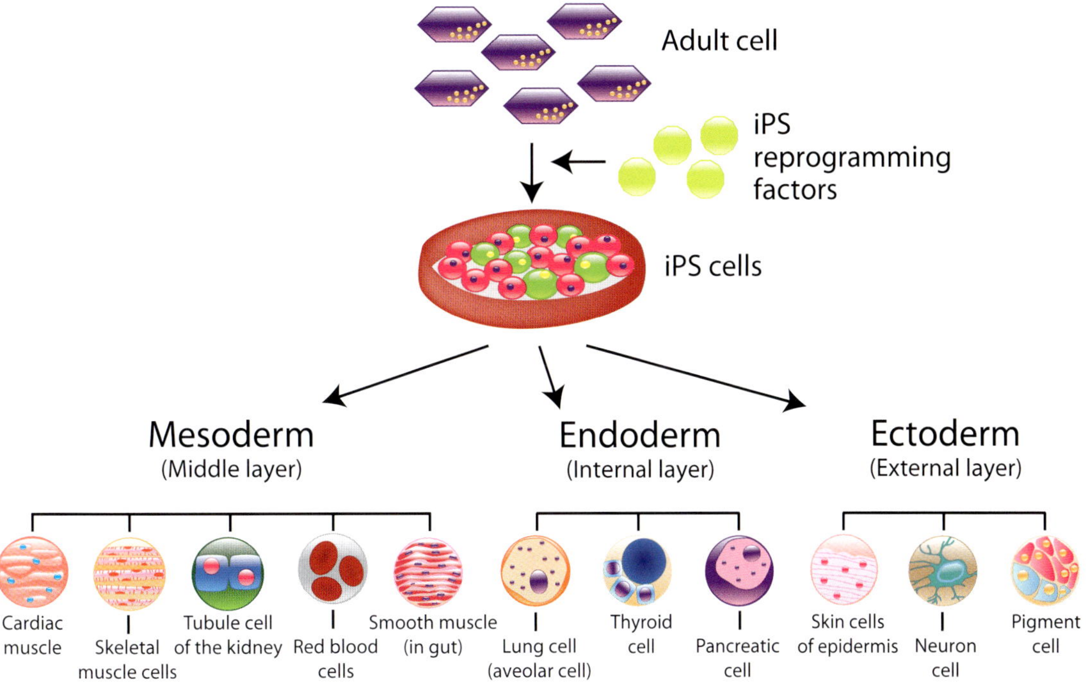

Figure 10.3 Diagram depicting the process of generating induced pluripotent stem (iPS) cells. These cells can differentiate into cells of all three germ layers (mesoderm, endoderm, and ectoderm).

tissues of the embryo and adult. In contrast, ESCs are maintained by forcing cells to self-renew and not differentiate. This would suggest that ESCs have been forced to exit the ICM developmental program. However, since ESCs can be reinjected into a blastocyst and very quickly behave like an ICM, ESCs clearly retain the *potential* to act as an ICM. Thus, ESCs retain the capacity for normal ICM development but are artificially forced to *suspend* this program in cell culture.

The developmental potential of ESCs is also demonstrated by their extraordinary differentiation capacity outside the context of a blastocyst. Normal ESC propagation in culture in a lab (i.e., in vitro) artificially forces ESCs to duplicate themselves (self-renewal). Although ESCs are propagated in vitro by forced self-renewal, they can be directed down specific developmental paths in the culture dish. Conversely, ESCs are *defined* by the ability to differentiate into cells of all three germ layers (the groups of cells formed during the development of an embryo)—endoderm, such as gut cells; ectoderm, such as skin cells; and mesoderm, such as bone cells (Figure 10.2). Thus, whether in normal or artificial cellular contexts, ESCs clearly retain the ability to generate a vast spectrum of differentiated tissue.

Reprogramming and Cloning

The normal developmental potential of adult stem cells (aSCs) is very limited in comparison to the potential of ESCs. However, experiments in the last few decades have uncovered remarkable plasticity of both aSCs and their mature progeny. While highly artificial in nature, these results nevertheless demonstrate that cellular plasticity is a bone fide phenomenon, given the right conditions.

The artificial forcing of adult cells (either stem cells or their progeny) or of the nuclei of adult cells back to an embryonic state is called *reprogramming*. When adult cells are reprogrammed back to the zygote stage, the process is termed *cloning*. Embryonic stem cells can be derived via both of these processes.

Cellular identity is dependent upon the specific genes that are turned on or off. Therefore, attempts to convert one cell type into another have involved artificially turning on cell-type-specific genes in cell types in which these genes are normally turned off. For example, it was hypothesized that adult cells could be forced back to ESCs since some of the genes controlling ESC self-renewal are known. Investigators successfully forced adult cells to return to the embryonic state by artificially expressing (turning on) just four embryonic genes. ESCs derived by reprogramming adult cells and aSCs are termed *induced pluripotent stem cells* or "iPS" cells (Figure 10.3).

Cloning is the process by which ESCs or a new individual are derived without knowing the specific signals involved in the reprogramming process. Technically, iPS cell derivation is also a form of cloning, but the term "cloning" is usually restricted to the following methodology. Rather than identify the specific proteins and genes that control the zygotic state, investigators use the egg itself as a tool to re-wire adult cell gene circuitry. Since eggs are derived by the process of meiosis (see Chapter 5), they possess only half of the DNA content of a healthy adult. (The sperm normally supplies the other half.) Therefore, to clone a cell or a person, the egg's nucleus—which contains the DNA, the genetic material—is removed by a process termed *enucleation* and replaced with the nucleus from an adult cell, restoring the full complement of genetic material to the egg. Furthermore, since the DNA of the egg is derived entirely from the donor nucleus, the DNA of the resultant clone exactly matches the DNA of the person donating the nucleus. An electric shock is applied to induce the egg to divide and act like a zygote. If successful, the new zygote will begin the process of early development and develop into a blastocyst.

There are at least two potential fates for a cloned blastocyst (Figure 10.4). If ESCs are desired from the cloned blastocyst, the blastocyst is destroyed and ESCs are harvested. This process is termed *therapeutic cloning* or *somatic cell nuclear transfer* (SCNT). If a newborn is desired, the cloned blastocyst is implanted into a human uterus. The implanted woman becomes pregnant and acts as a surrogate mother. This fate of the blastocyst is termed *reproductive cloning* and it results in a newborn genetically identical to the person donating the nucleus. These two individuals—the cloned one and the original donor—are analogous to identical twins. Thus, the difference between therapeutic cloning and reproductive cloning is not one of technique or biology, but of ultimate fate and purpose.

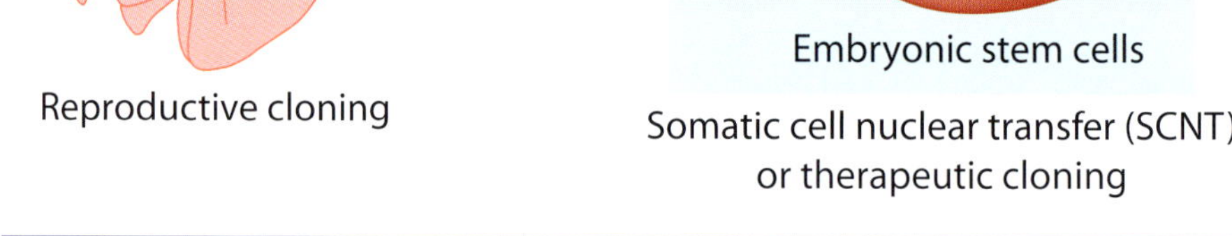

Figure 10.4 Diagram depicting the process of human cloning, resulting in either therapeutic or reproductive cloning.

The precise molecular signals within the cytoplasm of the egg that provoke reprogramming of the donor nucleus are unknown. However, these signals turn genes on and off rather than permanently change the DNA itself, since the same DNA is present in the zygote as well as in adult cells.

In summary, adult cells normally maintain a specific identity, but the processes of reprogramming and cloning can artificially force adult cells to assume very primitive embryonic states.

Stem Cells and Medicine: Cell Therapy

Before discussing the ethical implications of aSC and ESC use, it is necessary to highlight the current and theoretical therapeutic uses of these stem cell types. The polarized nature of the ethical discussion arises largely as a result of the therapeutic potential and limitations of these cells.

Adult stem cells have been used clinically for over four decades. For example, bone marrow transplantation, a form of blood stem cell transplantation (though unknown as such initially), has been a regular and important therapeutic regimen since the late 1960s. However, the same biological properties of aSCs discussed earlier—limited tissue distribution, limited developmental potential, and limited in vitro propagation potential—restrict the clinical use of these cells in human patient therapy. For example, no aSC treatment exists for degenerative brain diseases such as Parkinson's or Alzheimer's since the existence of aSCs in the brain is still unclear. Furthermore, since blood stem cells produce only blood cells and not brain cells, bone marrow transplantation is not a viable treatment option for these patients. Thus, despite lengthy clinical use, aSCs have yielded valuable but limited therapeutic fruit.

In contrast to aSCs, ESCs have had few, if any, documented clinical successes. However, this is partially due to a relatively limited opportunity for clinical exposure, not having been successfully isolated until 1998. ESCs overcome the limitations of aSCs, resulting in greater expectations for their usefulness. ESCs can be induced, theoretically, to produce any tissue of the body and can be propagated indefinitely. Because of these unique characteristics, it is likely just a matter of time before ESCs find wide clinical application.

However, as highlighted above, the directed differentiation of ESCs remains a large and difficult hurdle to overcome, and this hurdle *must* be solved before ESCs can be used clinically. In addition, ESCs form teratomas (a type of tumor) upon injection into a mouse. Therefore, it is dangerous and unwise to attempt to use ESCs clinically until protocols have been developed that promote the differentiation of ESCs with 100 percent efficiency. Thus, ESCs by themselves are not therapeutically useful. Rather, it is their cell products, aSCs that have been differentiated from ESCs, that may be clinically relevant.

Both aSC and ESC therapies are encumbered by a transplantation hurdle—the immune incompatibility between donor and host. In aSC therapy, this hurdle is partially overcome by employing autologous transplants (or "self" transplants) in which donor and host are the same. However, if the patient has cancer (for example, leukemia), this option is dangerous due to potential contamination of the transplant by cancer cells. Conversely, for ESC therapy, sperm and egg fusion results in the creation of a genetically unique individual with a unique immune system. ESC derivation from the resultant blastocyst yields ESCs with unique immune markers that are likely incompatible with the recipient patient. Thus, both aSC and ESC transplantation therapy stalls without some means to match donor and host immune systems.

As described above, the process of therapeutic or reproductive cloning results in a blastocyst that is genetically identical to the individual donating the nucleus to the egg. ESCs derived from this cloned blastocyst would circumvent the immune rejection problem. An individual requiring a cell transplant would simply clone himself or herself via SCNT and use the cloned ESCs to produce the cell type therapeutically required. If this were possible, ESCs would theoretically be far superior therapeutically to aSCs since ESCs would overcome all of the aSC limitations.

Currently, no kind of cloning, either therapeutic or reproductive, has been publicly reported in humans. The process of cloning Dolly the sheep required hundreds of attempts, and even after a single success, the resultant clone showed signs of premature aging and sickness. This result suggests that, even when successful, the cloning process using publicly available technologies is an aberration and does not fully re-wire the circuitry necessary to produce a "new" pure zygotic state.

Solving the cloning hurdles: iPS

As described in prior sections, the process of reprogramming, specifically the process of iPS generation, has been successfully performed. Technically, iPS technology is a form of cloning since the derived ESCs are genetically identical to the individual who donated the adult cells from which the iPS/ESCs were derived. In contrast to the traditional means of cloning that uses an egg (SCNT), iPS technology is not encumbered with the technical hurdles of SCNT. While the efficiency of iPS generation is low, it is much higher than that for SCNT. However, current iPS technology uses viruses to turn genes on. The viruses permanently alter the genome of the induced cell. Therefore, iPS cells are presently not useful in the clinic. These hurdles appear tractable and minor in comparison to those facing SCNT. Nevertheless, iPS results in ESCs and the issue remains of 100 percent efficient differentiation of ESCs into clinically relevant cells. Although iPS solves many (but not all) of the problems found in SCNT, it also raises new ones.

Theoretically, the most efficient and safest means to generate cells for therapy is partial reprogramming (Figure 10.5). Partial reprogramming is a process in which adult cells are artificially and directly differentiated into cell types different from those that would normally be produced. For example, partial reprogramming might be used to force a white blood cell to differentiate into a neuron. This technology seems promising since the iPS technology (a form of reprogramming) on which it is based has already been successfully accomplished. Furthermore, partial reprogramming would avoid the safety concerns encountered with inefficient ESC differentiation, since no tumor-producing ESCs would be generated at any step of the process. To date, preliminary reports of success have been published. Clinically relevant protocols await more research.

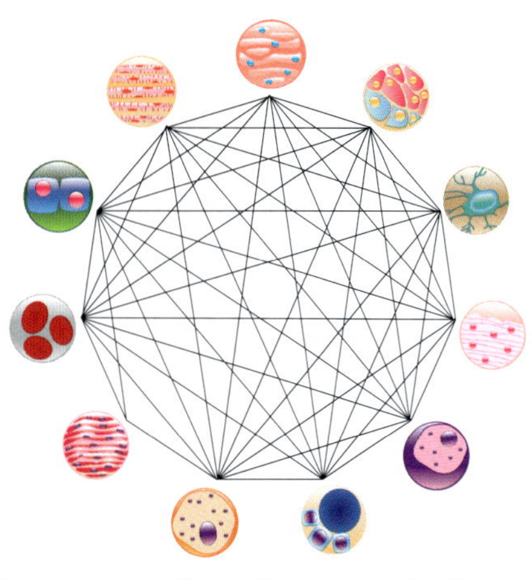

Figure 10.5 Partial reprogramming in which adult cells are artificially and directly differentiated into other cell types.

In summary, ESCs show far superior clinical potential compared to aSCs since they overcome the aSC limitations. However, ESCs have practical barriers to clinical use as well. Even though recent technological advances solve some of the issues, there still is no universal stem cell "silver bullet" for treating human disease.

Stem Cell Ethics

As the discussion above demonstrates, neither aSCs nor ESCs are the ideal solution to the problem of stem cell transplantation. The theoretical potential and power of ESCs has excited many in the scientific and medical community. However, the moral dilemmas raised by the use of stem cells need to be addressed.

Defining the question

There is virtually no opposition to the use of aSCs in medical research because aSCs are harvested from adults, and this does not require ending human life. The ethics of stem cell research revolve largely around ESCs. ESC harvest, cloning, and iPS generation involve the very early stages of human development, making ESC research extremely controversial.

The major question facing the ethics of ESC research is the definition of the beginning of human life. This is not a scientific question, however, but rather a moral one. Though science is an excellent tool by which we understand and describe human development, science is limited to describing cell divisions, stages, and behaviors. Science cannot define one stage or another as "life." This requires a moral judgment. Hence, it is entirely legitimate and necessary to invoke philosophy and/or religion in the context of ESC ethics.

The first source of moral authority that a Christian must consult is the Bible. Though Scripture does not mention ESCs, it does explicitly define when life begins. In Luke 1, when Mary greets Elisabeth after both women have been told that they would bear children, "the babe leaped in her [Elisabeth's] womb." Scripture does *not* say that "an extension of Elisabeth's uterus" or "a blob of tissue" leaped in her womb; rather, an independent entity ("the babe," which we know is John the Baptist) did the leaping. Thus, Scripture teaches that life exists before birth.

Furthermore, Scripture teaches that life begins at conception. In Psalm 51, written by David after his sin with Bathsheba, he confessed that "in sin did my mother conceive me." Note that David did *not* say "in sin did my mother conceive a blob of tissue that *became* me." Rather, David was "me," a unique sinful individual from the moment his mother conceived him. Thus, according to the Bible, life starts the moment that sperm and egg unite.

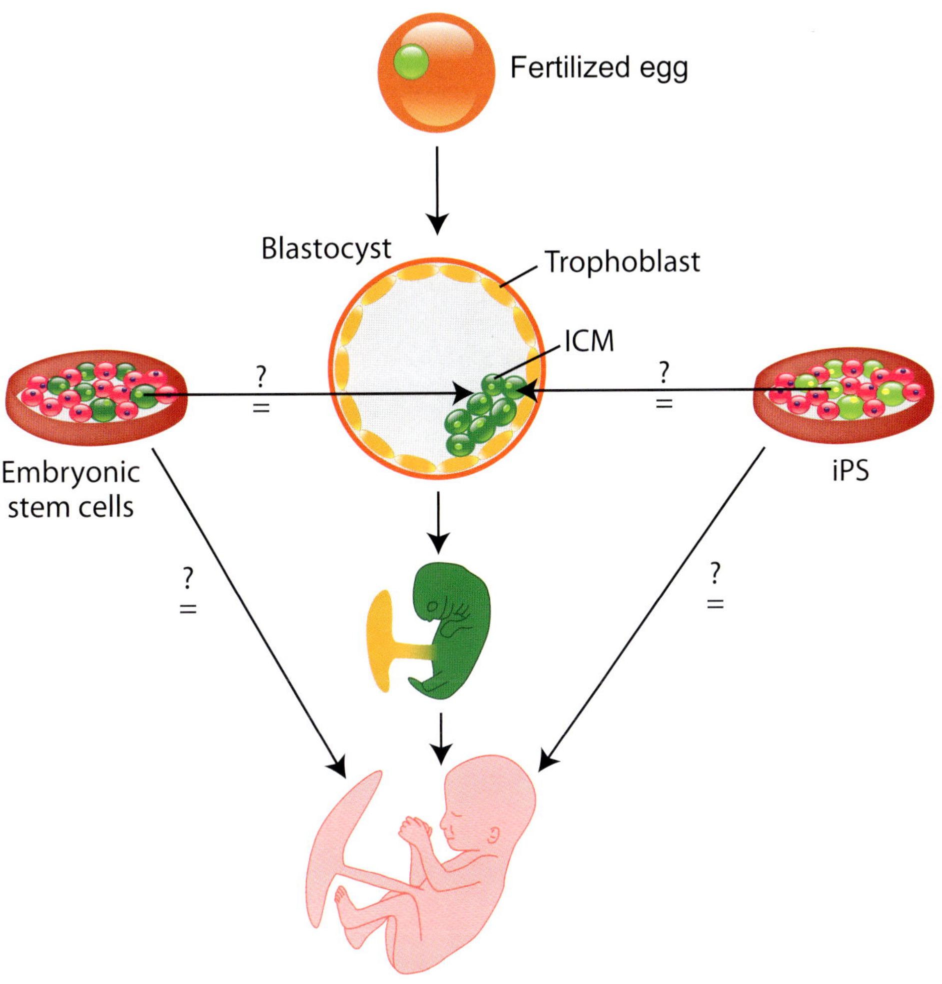

Figure 10.6 The ethical questions involved in research involving human ESC and iPS cells: Are ESCs equivalent to the ICM that, if uninterrupted, develops into a human fetus, and does the generation of iPS cells lead to the creation of new human life?

With these Scriptures in mind, it is now possible to frame the pressing ESC ethics question in a biblical light: Does ESC research and therapy take human life? If it does, then the ethics are clear: ESC use is morally unacceptable and wrong. If it does not, then there exists ethical freedom to use such cells.

Since life begins at conception, the first question relevant to ESC use is: Does ESC derivation harm, incapacitate, or destroy a zygote? Many would conclude that ESC derivation takes human life because it dismembers the blastocyst, a post-conception stage of human development. Indeed, former President George Bush's prohibition of the use of federal funds for new ESC derivation was based on this premise. If ESC derivation does indeed terminate development, then ESC derivation is immoral.

However, it is important to recall the functional equivalence between ESCs and ICM, as discussed in a previous section. If the ICM and ESCs are indeed equivalent, this would suggest that ESC derivation does not end human life, but may, rather, *suspend* human life indefinitely. Recall that, at the blastocyst stage, two cell populations are discernible—the trophoblast layer and the ICM. Since the ICM, not the trophoblast layer, eventually spawns the fetus, it seems appropriate to label the ICM as an early "fetus" and the isolation of ESCs as analogous to the removal of a fetus from the womb—specifically, from the placenta in the womb. However, unlike fetuses, ESCs can be propagated and theoretically rescued by re-implantation into an empty blastocyst, as has been done in mice. Nevertheless, since there is no source of empty human blastocysts into which ESCs can be injected, the isolation of ESCs *effectively* ends human life. Thus, whether ESC derivation is viewed as *termination* or *suspension* of human life, the ethical status of ESC isolation from blastocysts is the same immoral.

While this distinction between *termination* and *suspension* of human life may seem like a semantic distinction with little practical ethical consequence, it has enormous implications for the ethics of ESC derivation via iPS technology. For the derivation of ESCs via the iPS protocol, the *termination* line of reasoning sees no ethical dilemma, since no human life is created or destroyed in iPS technology. Thus, ESCs derived via an iPS protocol do not represent an actual fetus. In sharp contrast, the *suspension* line of reasoning sees the iPS protocol as leading to the creation of new human "fetuses"—albeit at an early stage, but new (cloned) human lives nonetheless. This *suspension* line of reasoning also concludes that the creation of new life by iPS is unacceptable since it also simultaneous leads to the irreversible suspension—effectively, to the destruction—of this new life due to the lack of means to rescue these new fetuses. Thus, deciding whether ESCs and the ICM are equivalent has enormous implications for the ethics of ESC use (Figure 10.6).

Though the morality of ESC research depends heavily on one's philosophical and religious persuasion, the disturbing nature of ESC derivation is evident independent of the Bible. ESCs can theoretically form any cell type of the body. This fact alone should be sufficient to give any researcher or clinician pause. No other tissue source yields as much developmental power as ESCs. Only the sacrifice of a healthy adult could yield such a broad spectrum of cell types and developmental possibilities. This apparent clinical and therapeutic equivalence between a healthy adult and an ESC line should cause any observer to stop and consider the moral status of ESCs. Hence, even from a non-scriptural perspective, the use of ESCs is morally troubling.

One potential resolution to the ethical dilemma of ESC use may come, not from ethical debate, but from technological advancement. Theoretically, the process of partial reprogramming would solve nearly all of the practical hurdles facing ESC and aSC therapy. Since this technology is designed to convert one adult cell type into another adult cell type without any embryonic intermediate, partial reprogramming would generate clinically useful cells without creating new human life or taking human life. The only ethical concern in this process would be ensuring that the reprogramming process never unintentionally converts the cells used to an embryonic state. If this concern is addressed and if partial reprogramming technology is realized, ethical discussions of stem cell use may very well be moot in the future.

Evaluating objections to the restriction of human ESC research

Not surprisingly, there are many, particularly in the scientific community, who disagree with the above conclusions and forcefully argue that ESC research is not only morally acceptable, but also morally imperative. When attempting to maintain a biblical perspective while evaluating such objections, it is often useful to replace the term "ESC" or "blastocyst" with "healthy human adult." Since the blastocyst is such an early stage of development, and since a hollow ball of cells may not "look like a human," some have difficulty consistently applying the biblical doctrine that life begins at conception. Conversely, the emotional repulsion one might normally feel upon learning of the slaughter of innocent victims is often weaker if the victims were embryonic rather than if the victims were healthy adults. Thus, since Scripture gives the blastocyst and the healthy adult the same moral status, equating the two mentally and conversationally may be a profitable exercise.

The first common objection to the prohibition of human ESC research is: "We don't know when life begins." The next logical step in this line of thinking is: "Therefore, we should go ahead and do ESC research." This objection cleverly shifts the burden of proof from those requesting permission to do research to those objecting to it. Clearly, in questions of morality, ignorance is a poor excuse for engaging in potentially immoral behavior. Would anyone permit researchers to dismember healthy adults if the researchers claimed ignorance on when life begins? Without question, appealing to ignorance is, at best, an excuse to *delay* the research, never to permit it.

The second major objection to restricting human ESC research is the pragmatic claim that ESC research "saves human life." In fact, the proponents of this position go so far as to claim that mankind has a *moral obligation* to pursue ESC research for the sake of the sick and diseased and that *prohibiting* human ESC research is *immoral*. Again, the analogy to healthy adults is illuminating: Would any sane human being view the slaughter of innocent, healthy human adults for the sake of body parts as a *moral imperative*? Clearly, this objection, like the first, also collapses upon close inspection.

The third ethical "excuse" for human ESC research involves embryos taken from fertility clinics. Unfortunately, the practice of in vitro fertilization often involves the creation of more embryos than the infertile couple intends to birth. There are clinics with freezers filled with "excess" or "leftover" embryos which either no one is aware of or no one desires to keep. Eventually, these embryos get discarded. Proponents of ESC research claim that since these embryos are going to be discarded anyway, why not make some good use of them? The rebuttal to this deceptive objection is persuasive once the analogy to the adult is used. Imagine a crime scene in which a criminal holds a gun to an innocent (healthy adult) victim's head. Would anyone in their right mind respond by saying, "Since he/she is going to die anyway, give me the femur and the heart when you've killed him!"? Clearly, only the morally perverse and cowardly would avoid the courageous act of attempting to rescue the victim. Conversely, the first response to the question of the fate of "leftover" embryos should not be "How can we harvest the cells for medicinal use?" but "How can we save these new lives from death?" Thus, this third excuse for ESC research also fails to be either persuasive or morally sound.

Despite the emptiness of the above objections, they are still commonly given in public discussions of ESC research. Thus, it is imperative for Christians to highlight the deficiencies in these objections and to present the clear and compelling biblical alternative.

For Christians in the public sphere, it is tempting to avoid citing the Bible in ethical discussions and resorting to pragmatic arguments against ESC research. A common example is the claim that aSC research has produced far more clinical successes than ESC research, and that ESC research is therefore unnecessary. While this claim is true at present, the discussion above demonstrates the limited applicability of this line of reasoning. Given that ESCs have been researched for far less time than aSCs and that ESCs theoretically overcome many of the aSC limitations, it is likely just a matter of time before ESC successes surpass aSC successes. Furthermore, as stated above, the question of when life begins is a moral and ethical one, not a scientific one; hence, citing the Bible is perfectly acceptable. Thus, it is unwise from both a therapy perspective and a philosophical perspective to resort to strictly pragmatic arguments against ESC research.

The complexity of the morality in the debate over ESC research is yet another example of the importance of maintaining a "Bible-first" perspective on all of life. When confronting thorny issues like ESC research or the creation/evolution debate, forsaking a Bible-first perspective can be the first step on a slippery slope toward compromise. Francis Collins, the well-known former head of the human genome project and outspoken atheist-turned-Christian, is also well-known for his compromise on the origins issue. He is an ardent theistic evolutionist, maintaining that God used evolution to create life and that Christians should stop doing "damage to faith" by insisting on the straightforward (Bible-first!) reading of Genesis. Dr. Collins is also in favor of embryonic stem cell research. As his example illustrates, relegating the Bible to second place in one area of life starts the slide down the slippery slope of compromise in other areas.

Conclusion

Adult stem cells sustain adult tissue turnover by self-renewing and by differentiating into the mature cells of the tissue in which they reside via unknown mechanisms. Due to their limited tissue distribution and to ignorance of the mechanisms controlling in vitro self-renewal, aSCs have limited therapeutic use. However, since they do not take any human life, aSC research and therapy are morally acceptable.

Embryonic stem cells are derived from the inner cell mass of the blastocyst and appear to be functionally equivalent to the ICM. Hence, ESCs differentiate into virtually any cell type, making them very attractive candidates for human patient therapy. However, this equivalence also entails that ESC derivation immediately or effectively terminates human life, making ESC research and therapy morally unacceptable. No form of ESC derivation, whether by SCNT or iPS technology, is without moral dilemmas, and the practical hurdle of efficient ESC differentiation into desired cell types further limits ESC use.

Thus, aSC research and therapy are the only unequivocally morally acceptable forms of stem cell research and therapy, and many of the practical limitations to aSC therapy may soon be solved by technological advancement, such as partial reprogramming.

Summary

1. *Embryonic stem cells* (ESCs) are derived from the *blastocyst* stage of early human development. This stage consists of the *trophoblast* layer surrounding the *inner cell mass* (ICM) from which ESCs are obtained. Removing the ICM destroys the blastocyst. Although ESCs are grown in the lab in a cell culture that allows them to self-renew but not differentiate, they still retain the ability to generate the full range of tissue cell types.
2. Adult cells can be forced back to an embryonic state in a process called *reprogramming*. When an adult cell is reprogrammed back to the zygote stage, it is called *cloning*. Adult cells that are reprogrammed by turning specific genes on or off are referred to as *induced pluripotent stem* (iPS) cells. Cloning usually involves removing the nucleus of an egg and replacing it with the nucleus of an adult cell. Both of these processes can be used to produce ESCs.
3. *Adult stem cells* (aSCs) have been used in human patient therapy for decades, but have limited applications. Because ESCs can be grown in the lab and can theoretically produce any type of tissue, they are thought to have much greater therapeutic potential than aSCs. Significant hurdles remain, however, to their clinical use. Recent technological advances have solved some of the issues, but an effective, problem-free method has yet to be developed for ESC treatments for human disease.
4. Since aSCs are harvested from adults, their use has generated virtually no opposition. The use of ESCs, however, has raised significant moral and ethical questions, the most important of which is whether embryonic stem cell therapy destroys human life. The Bible holds that life begins at conception. The question then becomes whether removing the ICM from a blastocyst *terminates* life or merely *suspends* it. The distinction has enormous ethical implications, but whether it involves termination or suspension, deriving ESCs from the ICM *effectively* ends human life and is therefore immoral. One possible solution to the use of ESCs may lie in future technological advancements, such as the partial reprogramming of adult cells in which no embryonic intermediates are needed.
5. The complexity of the moral debate regarding the use of embryonic stem cells in research or clinical therapy highlights the importance of maintaining a "Bible-first" perspective on all of life's issues. Abandoning or setting aside such a perspective leads to compromise and moral ambiguity. Science can describe the stages of human development, but it cannot define one stage as "life" to the exclusion of another. A moral judgment requires a moral foundation, and for Christians that foundation can *only* be the Word of God.

Cells: Sophisticated and God-Designed

Frank Sherwin, M.A.

God has created man with trillions of cells, each of which is a wonder of microminiaturization. The biochemical processes that occur moment by moment are nothing less than astounding. The bewildering complexity of these tiny functional units was acknowledged even before recent discoveries of how cells operate at the nanotechnological level.

> If you could build a motor one millionth of a millimetre across, you could fit a billion billion of them on a teaspoon. It seems incredible, but biological systems already use molecular motors on this scale.[1]

Indeed, "it seems incredible" to suggest that such sophistication is the result of chance, time, and genetic mistakes!

A 1997 *Nature* article by Steven Block detailed the "real engines of creation" that included a discussion of subcellular structures composed of springs, rotary joints, and levers—all made of protein.[2] The awareness of cellular sophistication has only increased by orders of magnitude since then, further demolishing the increasingly anemic Darwinian explanations for the origin of cells.

The cell or plasma membrane surrounding each cell has been called living because of its extremely precise selectivity—allowing or actively pumping some materials in or out, but not others. Complicated but efficient protein molecules "float" in the midst of this bilipid membrane. Some extend halfway and others all the way through the two interconnected membrane layers.

For the cell to remain alive, there must also be a constant exchange of materials from the outside of the cell to the inside, and vice versa. For example, among many other ions, potassium is critical for cellular function and homeostasis.[3] A precisely shaped and charged potassium gate found in the cell membrane is known to have a latch that rotates much like an iris! It also has switches and pulleys.[4] Working in exquisite harmony, the four principal parts of the gate—collectively called the Kir channel—are designed to selectively allow millions of potassium ions per second to pass through the gate while keeping out legions of pesky gatecrashers (other ions).

Cellular machines are not the stuff of randomness, but reveal unparalleled sophistication emanating from the mind of the wise Creator. Consider this quote describing the transfer of an electron to a heme portion of a ubiquitous protein involved in ATP (energy) production in living systems:

> This [electron] loading increases the redox potentials of both hemes a and a3, which allows electron equilibration between them at the same rate. Then, in 0.8 milliseconds, another proton is transferred from the inside to the heme a3/CuB center, and the electron is transferred to CuB. Finally, in 2.6 millisesconds, the preloaded proton is released from the pump site to the opposite side of the membrane.[5]

Is it logical to attribute such overwhelmingly complicated machinery to genetic mistakes "guided" by natural happenstance? No. Cellular research increasingly unveils amazing discoveries that should cause Darwinists to consider Paul's proclamation to the church in Rome: "For the invisible things of him from the creation of the world are clearly seen, being understood by the things that are made, even his eternal power and Godhead; so that they are without excuse" (Romans 1:20).

Sadly, for many biologists it doesn't matter what the burgeoning evidence shows. Those with a secular worldview must avoid pursuing the obvious design implication and therefore give glory to the creation instead of the Creator to whom it is due.

References

1. Feringa, B. L. 2000. Nanotechnology: In control of molecular motion. *Nature*. 408 (6809): 151-154.
2. Block, S. M. 1997. Real engines of creation. *Nature*. 386 (6622): 217-219.
3. Dubyak, G. R. 2004. Ion homeostasis, channels, and transporters. *Advances in Physiology Education*. 28 (4): 143-154.
4. Clark, et al. 2010. Domain Reorientation and Rotation of an Intracellular Assembly Regulate Conduction in Kir Potassium Channels. *Cell*. 141 (6): 1018-1029.
5. Belevich, I. et al. 2007. Exploring the proton pump mechanism of cytochrome c oxidase in real time. *Proceedings of the National Academy of Sciences*. 104 (8): 2685-2690.

Tendons, mostly made of connective tissue, attach muscles to bones. The darkly stained objects are cells that secrete long, fibrous proteins, seen as thin lines.

Appendix 1

Biology and the Bible

Henry M. Morris, Ph.D.

Biology, a word derived from two Greek words, *bios* ("life") and *logos* ("word"), is "the study of life." The Bible is the written Word of God, according to its own claims and an abundance of evidence.

The Bible encourages—in fact, commands—the study of biology and all other factual science. The very first divine commandment given to man was: "Be fruitful, and multiply, and replenish the earth, and subdue it: and have dominion over the fish of the sea, and over the fowl of the air, and over every living thing that moveth upon the earth" (Genesis 1:28).

This "dominion mandate," as it has been called, is in effect a command to "do science," for Adam and his descendants could only "subdue" the earth and "have dominion" over all its living creatures by learning their nature and functions. This clearly implies the establishment of a "science" of biology, so that mankind could properly care for and utilize the world's resources of animal and plant life as created by God.

There is thus no conflict at all between the Bible and biological science. But "evolutionary biology" is another matter. It is a philosophy, not science, an attempt to explain the origin and developmental history of all forms of life on a strictly naturalistic basis, without the intervention of special creation.

The Bible is opposed to evolutionary biology in that sense. Ten times in its opening chapter it stresses that the various created forms of life were to reproduce only "after their kinds" (see Genesis 1:11, 12, 21, 24, 25). This restriction does not preclude "variation," of course, since no two individuals of the same kind are ever exactly alike. Such "horizontal" recombinations, within the created kinds, are proper subjects of scientific study and so do not conflict with the Bible.

There are many fully credentialed professional biologists who are Christian creationists who have no problem with this biblical stipulation. The Institute for Creation Research, for example, has such professionals in the life sciences on its own staff, and there are hundreds more in other creationist organizations.

However, it is sadly true that *most* biologists and other life scientists are thoroughly committed to evolutionism. This is especially true of the biological "establishment." One poll of the members of the National Academy of Sciences found that, although commitment to atheism was predominant among the leading scientists in all fields, biologists were more so than others.

> Biologists had the lowest rate of belief (5.5% in God, 7.1% in immortality), with physicists and astronomers slightly higher (7.5% in God, 7.5% in immortality).[1]

In fact, probably most of this small minority who *do* believe in God are theistic evolutionists, not creationists.

However, it should be emphasized that this overwhelming commitment to evolutionism is not because of the scientific evidence, but rather because of antipathy to biblical Christianity. Even Charles Darwin became an evolutionist and agnostic because of his rejection of the biblical doctrine of divine punishment.[2]

Scientific evidence for biological evolution is very weak, at best. In all recorded history, there is no example of real evolution having occurred. The tremendous complexity of even the simplest forms of life is seemingly impossible to explain by evolution. Yet they believe it anyway. The genetic code which governs the reproduction process in all creatures is extremely complex, clearly implying intelligent design. Yet it is attributed to natural selection. Note the following statement.

> The genetic code is the product of early natural selection, not simply random, say scientists in Britain. Their analysis has shown it to be the best of more than a billion billion possible codes....Roughly 10^{20} genetic codes are possible, but the one nature actually uses was adopted as the standard more than 3.5 billion years ago.[3]

However, instead of coming to the obvious conclusion that an intelligent agent was responsible, it is simply assumed that it happened naturally.

> ...it is extremely unlikely that such an efficient code arose by chance—natural selection must have played a role.[4]

Natural selection thus takes the place of God, not only in the origin of species, but even in the origin of the remarkable code which governs life, so they say.

However, a number of evolutionary biologists have recognized the absurdity of relying on natural selection alone to accomplish such marvelous feats. Two very prominent evolutionists said it this way:

> Major questions posed by zoologists cannot be answered from inside the neo-Darwinian straitjacket. Such questions include, for example: "How do new structures arise in evolution?" "Why, given so much environmental change, is stasis so prevalent in evolution as seen in the fossil record?" "How did one group of organisms or set of molecules evolve from another?"[5]

These are the same unanswered questions that creationists have been posing to evolutionists for years. The obvious *true* answer is that of biblical creation.

This answer is not acceptable to evolutionists, of course, so they invent "just-so stories" or mysterious "order-out-of-chaos" scenarios.

> Fanciful abstractions have been invented by the neo-Darwinists, many of whom are scientists who, beginning as engineers, physicists, and mathematicians, found biology "easy."[6]

The coauthors of the book cited above, while vigorously opposing the neo-Darwinian concept of gradual evolution by random mutation and natural selection, are not endorsing the "punctuated equilibrium" hypothesis of Gould and others, and certainly not creationism. Rather, they think the answers lie in *Gaia*, the ancient pagan idea that the earth is a giant organism itself—Mother Earth, as it were.

Richard Dawkins is the best-known neo-Darwinist in England, with Edward O. Wilson (of Harvard) probably filling that role in America. A reviewer of one of Wilson's books noted that he "alludes in several passages to the problem of complexity as the greatest challenge facing all science."[7]

His co-Darwinian, Dawkins, thinks it can all be solved somehow in terms of computer simulations and his "blind watchmaker." However, in trying to explain the human brain by natural selection, Wilson seems to have come to an impasse.

> Evolution of the brain occurred over the three million years between our simian ancestors and the advent of *Homo sapiens* about a million years ago. The strangest feature of the process is that the capacity of the brain should far exceed the needs of mere survival. A further curiosity is that, once the brain was fully formed, the enormous differentiation of cultures occupied mere millennia, while only the twinkling of an evolutionary eye separates us from the earliest records of any civilization.[8]

Of course, none of this is strange or curious if one is willing to accept the biblical record of the origin of the human brain and the origin of civilization.

Instead of such a simple solution as primeval divine creation, however, evolutionary biologists argue violently among themselves about the relative merits of neo-Darwinism, punctuated equilibrium, and *Gaia* in explaining man. Stephen Jay Gould of Harvard (advocate of punctuationism) had a widely publicized debate with evolutionary anthropologist/linguist Steven Pinker. The comments by science writer Brookes are fascinating and relevant.

> Gould is an inevitable by-product of an age-old controversy which most scientists now acknowledge to be simplistic and well past its sell-by-date. It has no apparent function other than intellectual oneupmanship. It is precisely because there is so little evidence for either of their views that they can get away with so much speculation and disagreement.[9]

This particular debate was about evolutionary psychology, but the same comments could apply to evolutionary biology. Neither side can offer any observational evidence.

The punctuationists find their main evidence in the ubiquitous evolutionary gaps in the fossil record. In spite of these gaps, the fossil record is usually presented as evidence that evolution has occurred in the past, even though we cannot see it in either the field or lab in the present.

But the fossils don't really provide any solid evolutionary evidence either, whether for gradualism or punctuationism.

> Fossil discoveries can muddle our attempts to construct simple evolutionary trees—fossils from key periods are often not intermediates, but rather hodgepodges of defining features of many different groups....Generally, it seems that major groups are not assembled in a simple linear or progressive manner—new features are often "cut and pasted"—on different groups at different times.[10]

Not only are there no transitional series of fossils among the billions of known fossils in the rocks, but also there are no unequivocal evolutionary sequences.

> This poses a "chicken and egg" problem for paleontologists: If independent evolution of key characters is common, how is phylogeny to be recognized?[11]

The real bottom line of the entire question of biological origins is that the biblical record fits all the real scientific facts, and evolution does not.

References

1. Larson, E. J. and L. Witham, 1998. Leading Scientists Still Reject God. *Nature*. 394 (6691): 313.
2. Darwin, C. 1978. Autobiography. Reprinted in The Voyage of Charles Darwin. C. Rawlings, ed. BBC. See A Scientist's Thoughts on Religion, *New Scientist* (vol. 158, April 18, 1998), 15.
3. Knight, J. 1998. Top Translator. *New Scientist*. 158: 15.
4. Ibid.
5. Margulis, L. and D. Sagan. 1997. *Slanted Truths: Essays on Gaia, Symbiosis, and Evolution*. New York: Springer-Verlag, 100.
6. Ibid, 270.
7. Gillispie, C. C. 1998. E. O. Wilson's Consilience: A Noble Unifying Vision, Grandly Expressed, *Review of Consilience: the Unity of Knowledge* by Edward O. Wilson (New York: Alfred Knopf, 1998), 322 pp. In *American Scientist*. 86: 282.
8. Ibid, 281.
9. Brookes, M. 1998. May the Best Man Win. *New Scientist*. 158: 51.
10. Shubin, N. 1998. Vertebrate Palaeontology: Evolutionary Cut and Paste. *Nature*. 394 (6688): 12.
11. Ibid, 13.

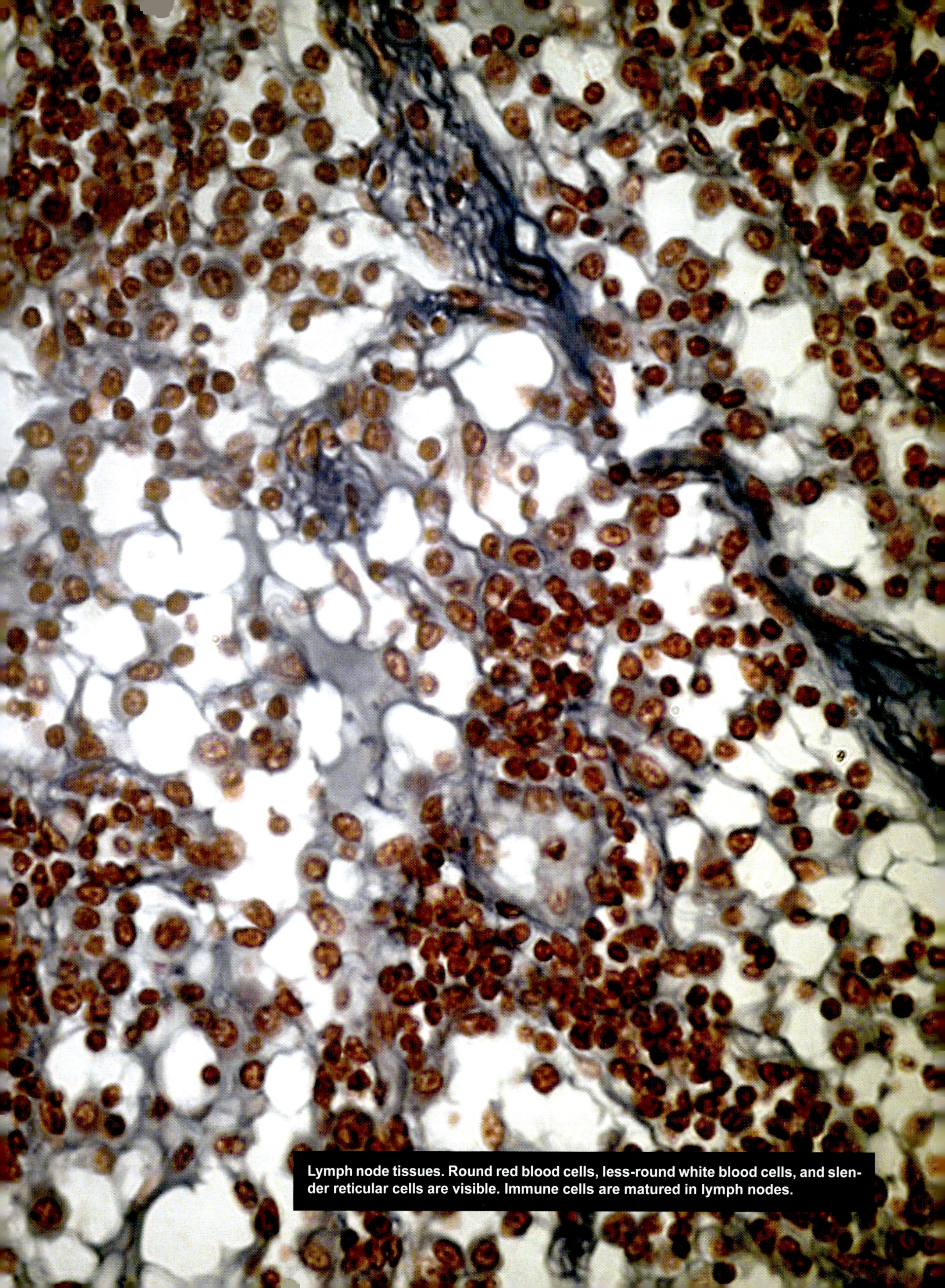

Lymph node tissues. Round red blood cells, less-round white blood cells, and slender reticular cells are visible. Immune cells are matured in lymph nodes.

Appendix 2
The Development of Pharmaceutical Therapeutics

Brad Forlow, Ph.D.

This book has exquisitely described the amazingly intricate design of the cell, the complexity of intracellular functions, and the remarkable cellular processes that give us life. Every function in the human body is accomplished through complex and precise molecular processes carried out by cells. These highly regulated and tightly controlled processes reveal an engineering and design marvel. However, at times abnormalities occur in the cellular processes of the human body. Underlying the common health problems discussed below is deficient or defective cellular functions or processes.

One of the more serious consequences of the Curse placed on creation due to Adam and Eve's rebellion in Eden is the prevalence of disease and physical and mental disorders. Thirty-four million Americans have been diagnosed with asthma during their lifetime. Heart disease is widespread in our society, affecting 27 million. One in three adults has high blood pressure, one in six adults has high cholesterol, and 26 million Americans live with diabetes, all risk factors for heart disease. Fifty million Americans suffer from some form of arthritis. Twenty-five percent of the population is affected by mental health illnesses, including 5.4 million with Alzheimer's and one in 20 with depression. Some form of cancer infects an estimated 12 million Americans.

These are not very uplifting statistics. Diseases and disorders like these significantly affect the quality of life of millions. Heart disease, cancer, cerebrovascular disease, respiratory diseases, diabetes mellitus, Alzheimer's disease, and primary hypertension are among the list of the 15 most common causes of death in the United States. In fact, heart disease and cancer are responsible for more than 50 percent of the annual death toll.

Therapeutic Intervention

Fortunately, we live in a time of great medical advances. Various medications are available for the treatment of many of these diseases and disorders, improving health and enhancing the quality of life. The most commonly used prescription medications include antidepressants, cholesterol-lowering drugs, and drugs to combat asthma, allergies, pain, acid reflux, high blood pressure, diabetes, and heart disease. In 2010, over $300 billion was spent by Americans on prescription medications.

Taking prescription medications has become a daily routine for many. A study in 2008 showed that 48 percent of Americans used at least one prescription drug, 31 percent used two or more prescription drugs, and 11 percent used five or more prescription drugs in the past month. Remarkably, 22.4 percent of children under the age of 12 take at least one prescription. Not surprising, prescription drug use increases as we age, with 29.9 percent of 12 to 19 year olds, 48.3 percent of 20 to 59 year olds, and 88.4 percent of people over 60 taking at least one prescription medication. These statistics may be startling, but for those who take prescription medication, can you imagine your life without the relief of symptoms that these medications provide daily?

Pharmaceutical Drug Discovery and Development

However, despite the common and widespread use of prescription drugs, most people are unaware of the extraordinarily complex and lengthy process of developing and launching a new medication. Although currently debated, it is reported that pharmaceutical and biotechnology companies spend as much as $1.8 billion to develop a new drug. It takes on average 10 to 15 years to create a new medicine that successfully achieves final marketing approval. Bringing a new drug to market is a multiphase process and requires a broad spectrum of expertise, including organic chemists, physiologists, biochemists, molecular biologists, in vivo pharmacologists, toxicologists, legal experts, physicians, regulatory specialists, statisticians, pharmacokineticists, computer scientists, IS/IT specialists, biologists, marketing expertise, project managers, and various engineers.

The high cost and extensive time requirement to bring a new drug to market is largely due to the extremely high attrition rate of drugs (Figure A.1). For every 5,000 to 10,000 drugs (chemical compounds) that are initially evaluated in drug discovery, only 250 enter preclinical evaluation. Only five drugs emerge from preclinical evaluation suitable for testing in human clinical trials. Remarkably, only one compound out of the five that qualified for human clinical trials is approved by the FDA to be prescribed by physicians for human use. The attrition of the drug candidates occurs at every stage of Drug Discovery and Development, mainly due to inadequate potency, inadequate selectivity, inadequate pharmacokinetic properties, preclinical and clinical safety (toxicity) concerns, and a lack of efficacy in the treatment of disease.

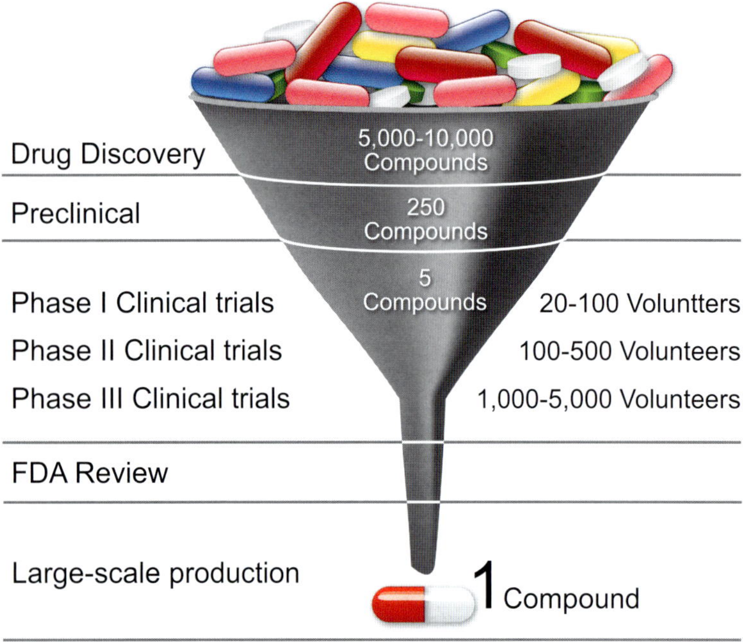

Figure A.1 Because of the many factors involved, new drugs require an extensive investment of time and money before they can be brought to market.

There are various stages of Drug Discovery (Preclinical R&D), Drug Development (Clinical R&D), and Drug Approval/Commercialization. Figure A.2 shows the distinct multi-step phases within Drug Discovery and Development. A potential new drug is evaluated at various stages within each phase. Embedded in the figure are the approximate time requirements for each phase of producing a new drug. Finding a safe and effective drug to combat human disease is a challenging venture. From basic science research to launching a product on the market, the multiphase processes of bringing a new drug from "bench to bedside" will likely astonish you.

Drug Discovery (Preclinical R&D)

Understanding disease pathophysiology: The body's immune system protects against infection by identifying and killing foreign substances (pathogens). The immune system is a highly complex and coordinated cascade of events involving a multitude of cell types, signaling pathways, enzymes, receptor molecules, and inflammatory mediators. A properly functioning immune response is vital for health and survival. A deficient or uncontrolled immune response underlies many of the diseases and disorders discussed above. In general, abnormalities in the immune response can be classified as either immunodeficiency disorders, autoimmune disorders, allergic disorders, or cancers of the immune system.

Immunodeficiency disorders result when a component of the immune system is absent or the immune response is diminished (e.g., IgA deficiency, Severe Combined Immunodeficiency (lack B and T lymphocytes), or leukocyte adhesion defects due to missing/defective components of the inflammatory response). Autoimmune disorders occur when the immune system mistakenly attacks and destroys healthy body tissue (e.g., multiple sclerosis, rheumatoid arthritis, systemic

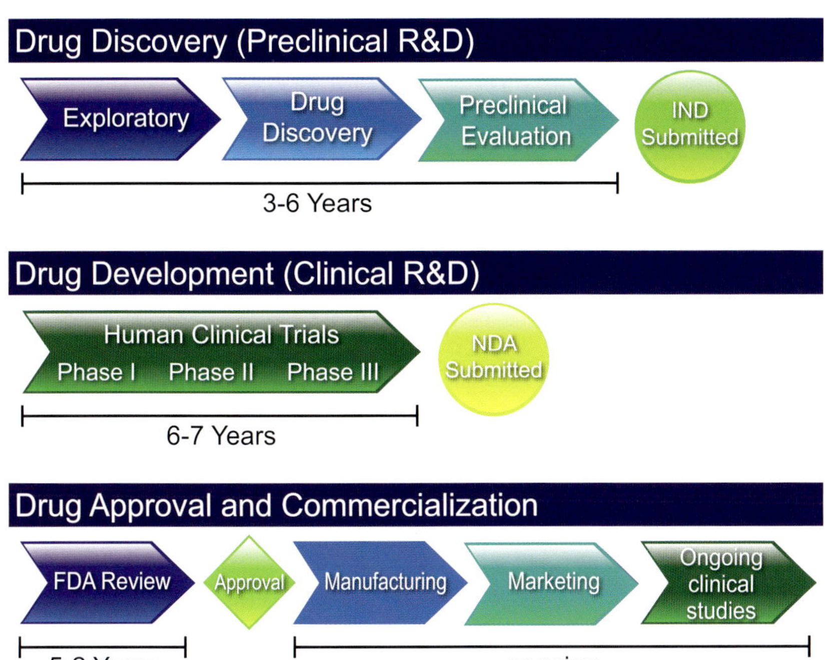

Figure A.2 The development of a drug involves multiple phases over a period of many years.

lupus erythematosus, and Type I diabetes). Allergic disorders result from a hyperactive immune response to allergens (antigens) in the environment, triggering an excessive inflammatory (inflammation) response (e.g., asthma, eczema, and environmental or seasonal allergies). Cancers of the immune system occur when cells grow out of control (e.g., lymphoma). There is strong evidence that inflammation, a significant component of the immune system, is involved in the pathogenesis of atherosclerosis (heart disease).

Due to the involvement of the immune response in the pathophysiology of many diseases and disorders, the immune response has been a key target for drug treatment, generally attempting to suppress an unwanted or overactive inflammatory response.

Figure A.3 depicts the various steps involved in this phase of a drug's development. Before a treatment (e.g., drug intervention) can be developed for a particular disease, the mechanisms involved in the disease at the molecular, cellular, and genetic levels must be elucidated. Whether considered pre-discovery, exploratory research, or basic science research, the pathophysiology of the disease or disorder must be studied. The cellular molecules, pathways, functions, or processes causative of or impacted by the disease or disorder (e.g., proteins, enzymes, receptors, or signaling pathways) must be discovered to provide a target for treatment.

Basic science research has provided invaluable insight into many human diseases at the molecular level. Recent advances in genomics and proteomics have attributed to this increased understanding of many diseases. An understanding of the cause of the disease and/or the pathways involved in the progression of the disease or disorder enables drug targets to emerge.

Exploratory
- Pathophysiology studies
- Target identification
- Target validation

Drug Discovery
- Lead compound identification
- Lead optimization

Preclinical Evaluation
- Preclinical safety and efficiency studies

Figure A.3 Diagram depicting the individual steps in the Drug Discovery phase.

Target identification: A target for potential drug intervention can be identified and selected based on an understanding of the cellular pathways (e.g., the intracellular signaling pathways, cellular processes, receptors, and/or proteins) involved in the development or progression of a disease or disorder.

A target is generally a single molecule directly involved in the cause or progression of a particular disease pathology (or associated pain) that can interact and be affected by a chemical compound or therapeutic agent. The interaction of the drug (or therapeutic agent) with the target produces a clinical effect through inhibition/blockade or activation of function. Common molecular and cellular targets for drugs include enzymes, receptors (e.g., ion channel receptors, G protein-coupled receptors, cytokine receptors, nuclear receptors, and thyroid hormone receptors), transport proteins (e.g., proton pump), nucleic acids (e.g., DNA and RNA), and cell-surface adhesion molecules.

Target validation: After a molecular target has been identified for drug intervention, the role of the target in the disease must be tested and confirmed. It is crucial to verify that the intended target is involved in the disease process. These experiments are conducted in both in vitro biochemical or cell-based assays and in vivo animal models of disease. Once the target for a disease or disorder is selected and validated, a safe and effective drug must be identified, developed, and commercialized in order to treat humans with the condition.

Lead compound identification: Prescription medications developed to provide therapeutic benefit to the patient can be generally classified as either small molecules or biologics. Most prescription drugs are small molecules. These are low-weight organic chemical compounds that bind with high affinity and specificity to the drug target (e.g., a protein or nucleic acid), either activating or inhibiting the function of the target. Genetically engineered living systems are often used to generate biological molecules as therapeutics (biologics). Certain targets are best inhibited by monoclonal antibodies (e.g., inhibition of growth factors or cytokines). Some diseases and disorders are treated with administration of a therapeutic protein (e.g., insulin or blood clotting factors). For the purposes of this discussion, the processes of drug discovery and drug development will focus on the use of small molecules.

After confidence in the target is established, the task is to find a small molecule (a "lead" compound) that could become a drug. Many times, high-throughput screening (HTS) is initially used to screen a company's vast library of chemical compounds for potential activity against a desired target. High-throughput screening is a highly automated and robotic process that allows hundreds of thousands of chemical compounds to be tested in a relatively short period of time (e.g., 100,000 compounds per day). HTS assays are designed to test for activity of the chemical compound against the target through the measurement of a biological response. Active compounds that modulate a particular biochemical or biomolecular pathway are considered "hits." From these assays, several lead compounds are identified (maybe upwards of 100) and selected for further study.

Compounds identified through HTS screening are tested in biological activity assays. The most promising compounds at this stage are assessed in safety studies (toxicology) and their pharmacokinetic properties are characterized (absorption, distribution, metabolism, and excretion—ADME). These studies are performed in cellular assays and animal models. Data from these experiments are analyzed to determine which drugs are likely to be best absorbed into the bloodstream, distributed to the proper site of action in the body, metabolized efficiently and effectively, successfully excreted from the body, and to be non-toxic. The lead candidates at this stage, potentially a few compounds from two distinct chemical classes, enter Lead Optimization to generate potent and safe compounds for preclinical evaluation.

Lead optimization: A chemical compound is desired that binds with high affinity and specificity to its target and affects biological function of the target, thus potentially impacting the disease process. Based on biological activity, pharmacological profile, and toxicology studies, the lead compounds are optimized to discover a more effective and safer compound. Structure-activity relationships (SAR) are used to direct chemists in the modification of the chemical structure of lead candidates. Slight modifications are made to the chemical compounds to improve activity against the target, to reduce activity against non-target molecules, and to improve drug characteristics (e.g., pharmacokinetic properties of absorption, distribution, metabolism, and excretion). These chemical synthesis efforts (combinatorial chemistry) result in hundreds of compounds (different variations or analogues) to be further characterized in selectivity (bind only to intended target and not to other molecules), potency, efficacy, toxicity, and bioavailability studies. The lead optimization studies are conducted in both in vitro cellular and biological activity assays and in animal models (in vivo).

Data from these experiments dictate further changes to the chemical structure that the chemists implement in the production of new analogues. At this stage, aspects of drug production must be considered, including formulation (e.g., can the compound be put into a form that can be dosed in animals and humans?), delivery mechanism (e.g., will the drug be taken orally, injected, or administered through an inhaler?), and large-scale manufacturing (e.g., can the drug be made in large quantities?). From these scientific evaluations and production considerations, potentially one to five candidates are identified that move into preclinical evaluation. In order for the lead compounds to be evaluated in preclinical studies, chemists must scale up their production to make the larger quantities required.

Preclinical evaluation: Rigorous preclinical evaluation is performed on one or more optimized compounds to determine which drug candidate(s) should be moved into human clinical trials. The primary hurdle in preclinical evaluation is safety, but efficacy in an accepted disease model is also required. Preclinical evaluation studies to determine safety and efficacy are performed in both in vitro assays and using in vivo models. Toxicology and in vivo pharmacokinetic studies are carried out in multiple species. From these studies, an estimation of the human dose is obtained. This guides the first large-scale production of the drug to determine if adequate quantities of the correctly formulated drug can be generated to conduct human clinical trials.

Before a drug can be administered to humans, it must be formulated for clinical use. Drug formulation involves producing the chemical compound in a manner that is safe for human consumption while maintaining the drug's chemical composition, purity, quality, and potency. The mode of delivery of the drug also impacts formulation for clinical use. For instance, the drug could be administered as a tablet, a time-release capsule, or a nasal spray.

Based on the safety profile, in vivo pharmacokinetics, efficacy in animal models of disease, and formulation data, it is determined whether a lead candidate (and a backup, at times) will be submitted to the Food and Drug Administration (FDA) for human clinical trials.

Drug Development (Clinical R&D)

At the end of the Drug Discovery process, researchers hope to have at least one candidate drug to test in people. A drug that has successfully traversed the stages of preclinical research and development enters Drug Development (Clinical R&D) upon approval by the FDA.

Investigational New Drug application: The FDA monitors and regulates the new drug development process. In order to conduct human clinical trials, an Investigational New Drug (IND) application must be filed with the FDA. These documents are extensive, thoroughly describing the chemical structure of the drug, its proposed mechanism of action in the body, the data from the preclinical evaluation studies, any side effects documented in toxicology studies, and how the drug is manufactured. The application must also include a detailed clinical trial plan that addresses each aspect of the clinical study, including how the study will be conducted, where the trial will take place, and the institution where the study will be performed.

Human clinical trials: Clinical trials are expensive to conduct; therefore, clinical trial design is of utmost importance to provide statistically meaningful data regarding pharmacokinetics, safety, and efficacy. Most trials today are classified as placebo-controlled, randomized, and double-blinded. In this study design, some clinical trial subjects receive the new drug candidate, but others receive a placebo. Randomized means that the clinical trial subjects are randomly assigned to either the drug candidate group or the placebo group. Double-blinded indicates that not only do the clinical trial subjects not know which treatment group they are in, but the researchers administering the drug do not know which treatment the subject is receiving. Traditionally, human clinical trials are conducted in three phases: Phase I, Phase II, and Phase III.

Phase I clinical trials are generally conducted in a small number of subjects, usually about 20 to 100 healthy volunteers, and typically take six to nine months to complete. In these studies, the subjects take the investigational drug for a short period of time. Generally, the drug is administered over one to two days to a week. Different groups of subjects receive increased doses of the drug, pending no safety issues arise in previous dose groups. Safety and tolerability of the drug in humans is the primary evaluation in Phase I clinical trials and are carefully monitored. What side effects were reported? Where there any adverse reactions? What is the maximum tolerated dose? However, the pharmacokinetics (PK) are analyzed to determine what the human body does to the drug (e.g., absorption, distribution, metabolism, and elimination). Pharmacodynamic (PD) studies are also performed to determine if the drug is producing the desired effect. PK/PD relationships obtained from the analysis of multiple blood/plasma samples collected over a period of time guide decisions regarding the drug dose range.

Phase II clinical trials are conducted in about 100 to 300 patients with the disease under investigation. Phase II trials generally take six months to three years to complete. In these studies, a multiple ascending dose is administered. These studies are conducted over a period of a few weeks, months, or a year. Safety is critically evaluated as short-term side effects (adverse events) and risks associated with the drug are recorded. Phase II studies are designed to not only study the safety of the investigational drug, but to determine the drug's effectiveness (efficacy) in patients suffering from the condition the new drug is intended to treat. Clinical endpoints are measured to determine if the disease is being impacted. PD effects are measured to determine the effective range of doses. Assays are performed to determine if the drug is altering the expected mechanism. The minimum effective dose and the maximum tolerated dose are established. Phase II may consist of only a few to twenty-plus trials.

Safety and efficacy are further investigated in a large group of patients in Phase III clinical studies, typically about 1,000 to 3,000 patients with the disease. Phase III clinical trials can take between one and four years to complete. Therefore, these trials are costly and long. Multiple clinical trial sites are used, both in the United States and around the world, to get a large and diverse group of patients. Phase III trials also include placebo, positive control, and active comparator groups. These studies provide confirmation of efficacy in the disease and an assessment of the risk/benefit. The complete safety profile is established, which also provides the basis for the labeling instructions (regulatory information).

Due to the high costs of clinical trials, additional phases have been added. Phase 0, Phase IIa, and Phase IIb studies attempt to reduce the costs and the length of clinical studies by acquiring information that could eliminate the drug candidate from evaluation in longer, more expensive trials. Small doses of drug are given to a small number of subjects in Phase 0 trials to determine human metabolic and biologic effectiveness. Phase IIa and Phase IIb studies provide the opportunity to obtain efficacy and dosage data alongside the safety data of a typical Phase I study.

Drug Approval and Commercialization

Submission of New Drug Application to FDA for approval: If the clinical trials reveal that the drug candidate is both safe and effective in treating the disease/disorder, the sponsoring company submits a New Drug Application (NDA) to the FDA for approval to market the drug. The application contains all the drug discovery (preclinical) and drug development (clinical) research data, including submission of all preclinical and clinical pharmacological, efficacy, and safety data. The submission also includes the drug's composition and the company's plan for manufacturing, producing, packaging, and labeling of the new drug. These applications can reach 100,000 pages! The FDA reviews all of the data submitted in the NDA to determine if the proposed drug is both safe and efficacious in the treatment of the human condition and if the proposed drug should be approved to be prescribed by physicians to humans.

The FDA review assesses the benefits and risks of the drug. Every drug produced has a risk (i.e., side effects); therefore, the FDA must determine if the benefits of the drug's effectiveness outweigh the risk. The FDA uses the benefit/risk assessment to determine the information that must be contained in the package insert (e.g., directions for use, potential side effects, and other necessary warnings stated in the package insert). This product labeling guides physicians in the use of the drug. The FDA is also concerned with the drug's manufacturing. FDA assesses the manufacturing methods to ensure quality that maintains the drug's composition, effectiveness, and purity.

Upon completion of NDA review, which can take 0.5 to two years, the FDA approves the medication for market, rejects approval, or requests additional studies before approval can be given.

Manufacturing: The complex process of bringing a new drug to market continues beyond human clinical trials and FDA approval. A new drug approved by the FDA for humans must be produced on a large scale. The manufacturing of the product to be prescribed is a highly regulated process that must meet strict FDA guidelines to ensure that the drug is safe for human consumption. Producing high-quality, large-scale quantities of the new drug compound requires manufacturing facilities, and each facility must follow Good Manufacturing Practices (GMP).

Ongoing clinical studies: Although bringing a new drug to market takes on average 10 to 15 years, research on the new drug continues after approval and marketing. Safety assessment of an approved drug is ongoing. Periodic reports are submitted to the FDA regarding adverse events. Phase IIIb/IV clinical studies are conducted for numerous reasons. The drug's safety and efficacy are assessed beyond the drug's original application. These studies enable the collection and analysis of long-term safety data on patients. At times, it is beneficial to test different dosage strengths and formulations. Also, many times new indications (new diseases/disorders) for the approved drug are investigated.

Pharmacological Inhibition

Despite the extensive challenges to bring a new drug to market, new medications are approved each year, bringing therapeutic relief to millions and enhancing their quality of life. For example, millions of Americans suffer from gastroesophageal reflux disease (GERD), high cholesterol, and allergies. Today, many treatment options are available to combat or resolve these abnormalities. The elucidation of the pathways of acid production, cholesterol production, and the allergic response provide excellent examples of how an understanding of the complex cellular pathways and processes involved in disease pathophysiology lead to viable targets for drug intervention. Medications developed for these conditions target very specific cellular pathways or processes critical in the pathophysiology of the disease/disorder, thereby providing tremendous therapeutic benefit while reducing the side effects.

GERD (acid reflux) or other conditions related to elevated gastric acid production affects millions of Americans. The cell types, proteins, and enzymes involved in the pathway of acid production have been elucidated. It was determined that the proton pump is the final step in gastric acid secretion, responsible for secreting H^+ ions into the gastric lumen. Therefore, the proton pump became an ideal target for drug intervention to inhibit acid secretion. Prilosec®, Prevacid®, Nexium®, and Protonix®, all small molecular compounds that block acid secretion, are commonly prescribed proton pump inhibitors for the treatment of GERD.

The active ingredient in Nexium® is bis(5-methoxy-2-[(S)-[(4-methoxy-3,5-dimethyl-2-pyridinyl)methyl]sulfinyl]-1H-benzimidazole-1-yl) magnesium trihydrate, a chemical compound with the molecular formula $(C_{17}H_{18}N_3O_3S)_2Mg$ x 3 H_2O (Figure A.4). Nexium® is a proton pump inhibitor that suppresses gastric acid secretion by specific inhibition of the H+K+-ATPase (hydrogen/potassium adensosine triphosphatase enzyme system) in the gastric parietal cell. By acting specifically on the proton pump, Nexium® blocks the final step in acid production, thus reducing gastric acidity.

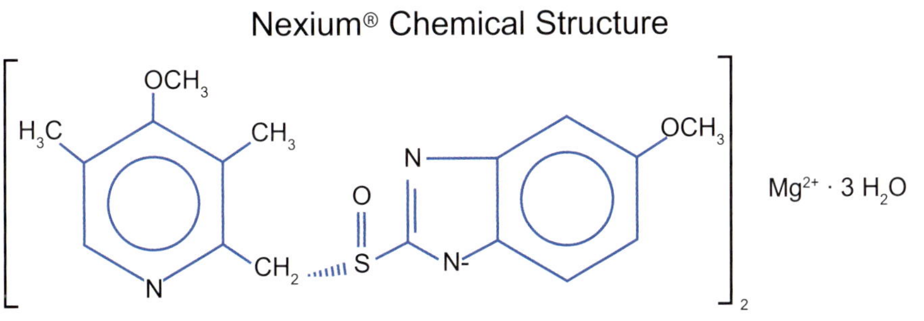

High cholesterol, a risk factor for heart disease, also affects millions of Americans. Cholesterol production is the result of a complex cascade of intracellular events. Figure A.5 shows a simplified diagram of some of the intermediates and enzymes involved in the production of cholesterol. The elucidation of this pathway provided many potential targets for the inhibition of cholesterol production. HMG-CoA is a target that has been exploited by many companies. Statins are a class of drugs used to lower cholesterol levels by competitively inhibiting the enzyme HMG-CoA reductase, which is involved in production of cholesterol in the liver. Multiple drugs have targeted this pathway, including Lipitor®, Zocor®, Crestor®, and Pravachol®. These reduce the production rate of mevalonate, blocking the pathway for

Nexium® Chemical Structure

$Mg^{2+} \cdot 3\ H_2O$

Figure A.4 The chemical structure of the drug Nexium®, which treats acid reflux disorders by blocking acid secretion.

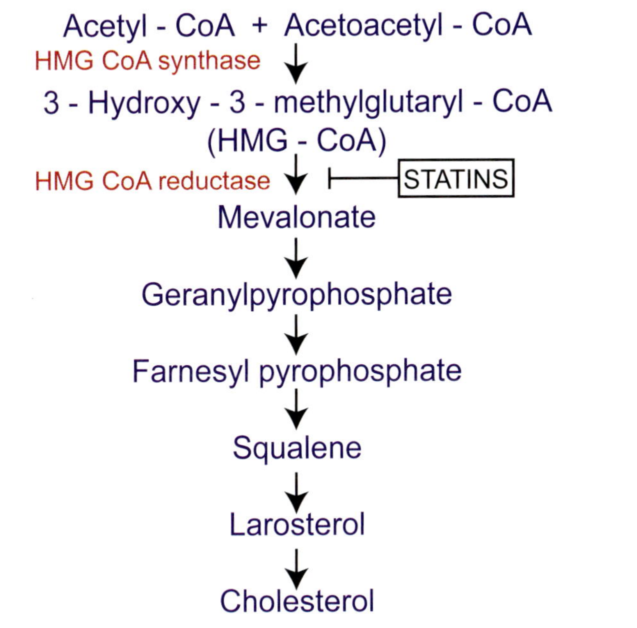

Acetyl - CoA + Acetoacetyl - CoA
HMG CoA synthase
3 - Hydroxy - 3 - methylglutaryl - CoA
(HMG - CoA)
HMG CoA reductase — STATINS
Mevalonate
Geranylpyrophosphate
Farnesyl pyrophosphate
Squalene
Larosterol
Cholesterol

Figure A.5 Cholesterol is produced through a series of cellular interactions. Statin drugs target the enzyme HMG-CoA reductase to block the metabolic pathway that leads to cholesterol synthesis in the liver.

cholesterol synthesis in the liver.

Allergies or asthma affect 50 million Americans. An allergic reaction is the result of a hyperactive immune response to an allergen. An allergen is a substance that is foreign to the body and can cause or "trigger" an allergic reaction. The exposure of the body to an allergen leads to an inappropriate/uncontrolled inflammatory response. Common allergens include dust mite excretion, pet dander, molds, and pollens (grass, weeds, and trees). Symptoms of allergies include rash, itchy/watery eyes, congestion, difficulty breathing, sneezing, and runny nose.

The molecular mechanism and cellular pathways of an allergic response are now well known. An allergen stimulates an antigen-presenting cell, which triggers B cells to produce antibodies (IgE). IgE antibodies stimulate mast cells to release various inflammatory mediators, including cytokines, histamine, leukotrienes, and prostaglandins. Histamine is a key component of an allergic reaction. Histamine causes local vasodilation, bronchoconstriction, and upregulates the inflammatory response (inflammation), inducing the release of acute inflammation mediators that affect neutrophils, eosinophils, basophils, monocytes, and lymphocytes. The allergic reaction propagated by an excessive inflammatory response is characterized by blood vessel dilation, bronchoconstriction, release of inflammatory mediators and cytokines, and the recruitment of inflammatory cells. Asthma, eczema, hay fever, and anaphylaxis are all clinical presentations of a hyperactive immune response to an allergen.

Histamine mediates its effect through binding to its receptors (H_1, H_2, H_3, and H_4) expressed on various cell types. H_1 receptor is expressed on smooth muscle cells, epithelial cells, endothelial cells, neutrophils, eosinophils, monocyte DC, T cells, and B cells. H_2 receptor is expressed on vascular smooth muscle cells, epithelial cells, endothelial cells, neutrophils, eosinophils, T cells, B cells, and mast cells. H_3 and H_4 histamine receptors are largely found in the central nervous system and in the bone marrow, respectively.

Due to histamine's central role in allergic reaction, histamine has become a viable target for drug intervention. Many people find relief to their symptoms of allergies in prescription medications (many of which are now available over the counter). Allegra®, Zyrtec®, and Claritin® are all extensively used allergy medications that target the histamine pathway by blocking the H_1 receptor.

Zyrtec® (cetirizine hydrochloride), available as a tablet or syrup, is a chemical compound with the formula of $C_{21}H_{25}ClN_2O_3 \bullet 2HCl$. Claritin® (Loratadine), sold as a tablet, syrup, or rapidly disintegrating tablet, has an empirical formula of $C_{22}H_{23}ClN_2O_2$. Allegra® is a chemical compound (Fexofenadine hydrochloride; $C_{32}H_{39}NO_4 \bullet HCl$) available as a tablet, an orally disintegrating tablet, or an oral suspension. Each of these chemical compounds is an active and selective H_1 receptor antagonist. Blocking H_1 receptors inhibits histamine activity, thus preventing the downstream allergic reaction stimulated by histamine production.

Conclusion

New targets continue to emerge from an increased understanding of the basic cellular pathways and processes related to diseases and disorders. Advances in science and technology have not only impacted basis science research, but also drug discovery and development. Pharmaceutical companies are devising new strategies to more efficiently (less time, less cost) conduct the appropriate clinical trials. Despite these advances and attempts to lower the high attrition rates of potential drugs, drug discovery and development remains an expensive and tremendously difficult process. Don't take for granted the medicines that you reach for daily to bring therapeutic relief, thereby enhancing your quality of life. For those who battle a disease or disorder, can you imagine your life without the benefits that these medications provide daily?

Glossary

Abiogenesis: The study of how life could have developed from nonliving matter through random natural processes.

Actins: Proteins that are involved in forming cell and tissue structure, shape, and motility.

Adenosine triphosphate (ATP): The primary energy molecule used to power various cellular activities.

Arabidopsis thaliana: A weedy, mustard-like plant that is the premier model dicot plant species.

Base-pairing: A form of complementary hydrogen bonding between nucleotide molecules in DNA strands that is partly responsible for the double-stranded nature of DNA.

Biofilms: Adhesive chemical films produced by colonies of bacteria and other microorganisms that cling to various types of surfaces. They provide support structures, communication, sustenance, and other features critical to communal life.

Blastocyst: The hollow ball stage of early embryonic development, consisting of the trophoblast layer and the inner cell mass.

Caenorhabditis elegans: A species of nematode, a worm-like creature used as a multicellular model system for research.

Capsule: The cell wall of a bacterium. Made of polysaccharides, it adds rigidity, structure, protection, and support.

Cell surface receptors: Molecules (protein complexes) imbedded in the plasma membrane of cells that bind signaling molecules of different types coming from neighboring cells or other parts of the organism.

Cell wall: Carbohydrate-based materials that form a strong wall surrounding the plasma membrane of a plant or bacterium cell.

Centromere: A specialized region of a chromosome that serves as a key attachment point for the machinery that separates DNA strands after replication prior to forming two new cells.

Chloroplast: An energy-related organelle primarily found in the cells of plants that converts water, carbon dioxide, and sunlight into carbohydrates that store energy through photosynthesis.

Chromatin: A complex, information-rich mix of DNA, protein, and RNA that comprises the chromosomes in the nucleus.

Codon: Three adjacent nucleotides in a messenger RNA that specify a certain amino acid to be added to a protein chain.

Cristae: The folds in the inner membrane of a mitochondrion that house and facilitate the enzymes that perform the reactions needed to produce a eukaryotic cell's energy.

Cytoplasm: The aqueous matrix inside a cell that houses its various organelles, molecules, and biological reactions and pathways.

DNA polymerases: Protein complexes that copy DNA and make new chromosomal strands for cell division. Some DNA polymerases also edit errors and repair breaks in DNA strands.

DNA watermarks: Man-made messages put into synthesized DNA strands that may then be inserted into the genome of a living organism.

Drosophila melanogaster: The common fruit fly, a frequent model cell system used in genetic testing.

Endoplasmic reticulum (ER): A network of membranes within the cell associated with and surrounding the nucleus. Rough ER is spotted with ribosomes for protein production, while smooth ER is not.

Enzyme-coupled receptors: Receptors that exhibit enzymatic capacity when stimulated by a signaling molecule. They are typically phosphatases or kinases that add or remove phosphate groups from associated proteins for the transmission of signal cascades inside cells.

***Escherichia coli* (*E. coli*):** A bacterium model system that also forms part of the complicated intestinal microbe community that enables the proper digestion of food and maintenance of the immune system.

Euchromatin: The regions of chromosomes that contain a higher number of protein-coding genes.

Eukaryotes: Single-cell or multicellular organisms whose cells contain distinct features such as a nucleus and organelles.

Exons: Regions within protein-coding genes that directly code for amino acids in the final protein product.

Extremophiles: Mostly single-cell prokaryotic organisms that live in extremely harsh environments.

Fibroblasts: Classes of cells that form a fibrous matrix in which cells are imbedded.

Filopodia: See lamellipodia.

Flagellum: A rotating, whip-like structure used by a cell to propel itself through its environment.

Formins: Key proteins involved in the recruitment of actin to form actin polymers that facilitate cell motility and activity.

G protein-coupled receptors: A large and diverse group of receptors that modulate cell processes in response to signaling molecules or environmental signals.

Golgi apparatus: A multifunctional organelle in plant and animal cells that processes large molecules such as proteins and lipids.

Guanine triphosphate (GTP): An energy molecule used in the cell that is similar to ATP.

Heterochromatin: Regions of the genome with various types of repeated sequences in which protein-coding gene activity is thought to be negligible.

***Homo sapiens* (humans):** Humans are unique because they are created in the image of God. There is only one human kind descended from one original ancestral male and female (Adam and Eve).

Homologous recombination: The process in meiosis in which similar chromosomes (maternal and paternal) pair up and exchange pieces of DNA from the same genomic regions.

In vivo: Taking place within a living organism.

In vitro: Taking place in a lab environment or culture dish instead of in a living organism.

Induced pluripotent stem cells (iPS): Adult cells that have been artificially converted back to an embryonic stem cell-like state.

Inner cell mass (ICM): The early cells of a new individual that eventually give rise to the embryo proper. They are located inside the hollow ball of the blastocyst stage.

Introns: Regions within protein-coding genes in eukaryotes that do not directly code for amino acids. Many have been found to code for regulatory RNAs, and they provide other key control features for proper gene expression.

Ion channel-coupled cell receptors: A class of cell surface receptors that help modulate an organism's bioelectrical system by controlling electrochemical impulses among cells through the flow of charged particles called ions.

Lamellipodia and filopodia: Lamellipodia are actin-filled, actively extended areas on the cell surface that are involved in cell movement. Filopodia are finger-like extensions from the lamellipodia region.

Lysosomes: Organelles that break down debris, waste products, and even bacteria and viruses that have been taken out by the immune system.

Meiosis: The process whereby the genome of a reproductive cell is copied, regions of maternal and paternal chromosomes are randomly exchanged (shuffled), and the chromosomes are eventually randomly sorted to new daughter cells that only have a single genome equivalent (23 chromosomes), half of the DNA complement of a normal body cell.

Mitochondria: Small organelles that produce the energy needed to power a wide variety of cellular activities.

Mitosis: The process whereby the genome of a cell is copied and separated into two new daughter cells, each produced by the dividing of a single parent cell.

***Mus musculus* (mouse):** A small animal used as a model for human gene and tissue studies because its genome is similar in size to that of humans and has many of the same standard genes.

***Mycoplasma genitalium*:** A single-cell microbe with one of the smallest gene complements known. This was one of the first microbial genomes sequenced because scientists wanted to develop a basic and minimal set of genes that were essential to life.

Nucleolus: A distinct region within the nucleus primarily containing ribosomal RNA genes.

Nucleus: A complex organelle in the cells of eukaryotes that contains the genomic material—chromosomes in the form of carefully constructed DNA and other important physical features.

Operons: Sets of genes in a single block that are under similar control features and typically transcribe as a group when activated. This is a common feature of prokaryote genomes.

Peroxisomes: Organelles found in most living cells that are critical to the metabolism of fatty acids in various pathways.

Phagocytosis: Process whereby the body's cells engulf food, debris, and even targets of the immune system (bacteria and viruses) so they can be broken down and recycled.

Photosynthesis: A complex set of photo-driven chemical reactions that occur in plant chloroplasts and convert water, carbon dioxide, and sunlight into storable energy.

Pili: Conduit-like tubes that bacteria form to exchange bits of DNA between themselves.

Plasma membrane: The outer boundary of a cell, consisting of a bilayer lipid membrane system that forms a selective barrier.

Prokaryotes: Single-cell organisms that do not have organelles or a nucleus.

Proteinoids: Globular, ring-like clusters of simple amino acids derived from highly contrived and engineered experiments for the purpose of finding answers to the impossibility of abiogenesis.

Racemic: An equal mix of right and left-handed molecules. Many small molecules, such as amino acids, can exist in both right and left-handed forms, yet biological systems only use the left-handed form to make proteins.

***Saccharomyces cerevisiae*:** A common strain of yeast and a single-cell eukaryote that has been turned into a model system like *E. coli* and studied extensively.

Synapses: Chemically active spaces that relay signals between nerve cells.

Translation: The production of a protein by polymerizing amino acids at ribosomes.

Trophoblast cells: The cells that are responsible for implantation of the new human into the mother's womb. These cells give rise to the placenta.

Tubulins: Structural proteins that are involved in forming and controlling microtubules within cells.

Vacuole: A large space within a cell enclosed by a membrane that can be used as a storage area, a site for degradation reactions facilitated by lysosomes, or other purposes.

Zygote: The first cell of a new individual. It is the product of the fertilization of the egg by the sperm.

Credits

Images

The Broad Institute of MIT and Harvard: Figure 2.2

Fotolia: Pages 21, 47, 99

ICR, Brian Thomas: 36, 82

iStock: Figures 3.1, 3.5, 8.2; pages 35, 65, 73, 83, 89

Dennis Kunkel (copyright © Dennis Kunkel Microscopy, Inc.): Figures 2.4, 3.4, 3.7, 3.9, 5.4, 6.4, 7.2, 7.4

National Institute of General Medical Sciences: Figure 5.2

National Institutes of Health: Figure 7.1

National Park Service: 2.5

Photo Researchers: Figure 5.5

Public domain: Figures 2.1, 4.4

Rocky Mountain Laboratories, NIAID, NIH: Figure 8.3

Travis Whitfill: Pages 12, 22, 48, 56, 60, 66, 74, 90, 100, 112, 116

Susan Windsor: Figures 1.1, 1.2, 1.3, 2.3, 2.6, 2.7, 2.8, 2.9, 2.10, 2.11, 3.2, 3.6, 3.8, 3.10, 4.1, 4.2, 4.3, 4.5, 5.1, 5.3, 5.6, 6.1, 6.2, 6.3, 7.3, 7.5, 7.6, 9.1, 9.2, 9.3, 9.4, 10.1, 10.2, 10.3, 10.4, 10.5, 10.6, A.1, A.2, A.3, A.4, A.5

Adapted Images

Figure 1.1 adapted from an image by Dan Cojocari, University of Toronto.

Figure 2.3 adapted from en.wikipedia.org/wiki/File:Cell_membrane_detailed_diagram_4.svg.

Figure 2.9 adapted from Jernigan, J. A. 2001. Bioterrorism-Related Inhalational Anthrax: The First 10 Cases Reported in the United States. *Emerging Infectious Diseases*. 7 (6).

Figure 4.1 adapted from an image by Madeleine Price Ball.

Figure 6.1 adapted from a Magnus Manske 2006 image.

Figure 9.2 adapted from Wilson, A. and A. Trumpp. 2006. Bone-marrow haematopoietic-stem-cell niches. *Nature Reviews Immunology*. 6 (2): 93-106.

Figure 10.1 adapted from Bailey, R. Stem Cell Disappointment – Culture War to Reignite? *Reason*. Posted on reason.com April 16, 2010.

Figures 10.2 and 10.3 adapted from Embryonic and Induced Pluripotent Stem Cells. Sigma-Aldrich Fact Sheet. Posted on sigmaaldrich.com.

Figure A.1 adapted from Kiyatec & Drug Discovery. Kiyatec Fact Sheet. Posted on kiyatec.com.

Adapted Articles

Thomas, B. Have Scientists Created a Synthetic Cell? *ICR News*. Posted on icr.org May 27, 2010.

Guliuzza, R. 2010. Life's Indispensable Microscopic Machines. *Acts & Facts*. 39 (8): 10-11.

Guliuzza, R. 2009. Made in His Image: Life-Giving Blood. *Acts & Facts*. 38 (9): 10-11.

Thomas, B. Where Did Flesh-Eating Bacteria Come From? *ICR News*. Posted on icr.org December 15, 2008.

Thomas, B. Brain's Complexity "Is Beyond Anything Imagined." *ICR News*. Posted on icr.org January 17, 2011.

Guliuzza, R. 2009. Made in His Image: Human Reproduction. *Acts & Facts*. 38 (1): 14.

Thomas, B. The Mysteries of Stunning Soft Tissue Fossil Finds. *ICR News*. Posted on icr.org December 20, 2010.

Guliuzza, R. 2009. Made in His Image: Immune Systems, the Body's Security Force. *Acts & Facts*. 38 (11): 10-11.

Guliuzza, R. J. 2009. Darwinian Medicine: A Prescription for Failure. *Acts & Facts*. 38 (2): 32.

Sherwin, F. 2010. Cells: Sophisticated and God-Designed. *Acts & Facts*. 39 (8): 17.

Morris, H. 2010. Biology and the Bible. *Acts & Facts*. 39 (11): 4-5.

Index

abiogenesis, 13-15, 17-19

actin, 75-77, 80

adenosine triphosphate (ATP), 40-41, 45, 68, 79, 111

adult stem cell (aSC), 94-96, 105-106, 108-109

Arabidopsis thaliana, 37

autosomes, 39, 61

base-pairing, 15, 49

biofilm, 25

blastocyst, 96, 101-105, 107-110

Caenorhabditis elegans, 37, 125

capsule, 28, 121

cell surface receptor, 67-68

cell wall, 28, 79

centromere, 59, 63

chloroplast, 41-42, 76, 80

chromatin, 39, 125

chromosome, 18, 29, 38-42, 45

cloning, 10, 20, 29-30, 103-106, 110

codon, 51-53

cristae, 40

cytokinesis, 57-58, 60, 63, 79

cytoplasm, 20, 24-25, 27-29, 40-42, 50, 52, 60, 68, 72, 78, 104

cytoskeleton, 25, 75, 79-80

DNA polymerases, 58

DNA watermarks, 19

Drosophila melanogaster, 37

embryonic stem cell (ESC), 91, 95-96, 101-103, 105-110

endoplasmic reticulum (ER), 43-44, 50, 52, 60, 79

enzyme-coupled receptor, 68-69

Escherichia coli (*E. coli*), 18, 26, 29-31, 33, 37, 51

euchromatin, 39

eukaryote, 30, 37-38, 40-41, 44-45, 50, 75

exon, 51

extremophile, 26

fibroblast, 78-80, 91

filopodia, 77

flagellum, 28-29, 33, 44-45

formin, 77

gametogenesis, 60-61

Golgi apparatus, 42-45, 53

G protein-coupled receptor, 68, 119

granum, 41

guanine triphosphate (GTP), 68-69, 76

heterochromatin, 39

homologous recombination, 63

Homo sapiens, 37, 114

induced pluripotent stem cells (iPS), 103-107, 109-110

inner cell mass (ICM), 96, 101-103, 107-110

intron, 52

ion channel-coupled cell receptor, 67

irreducible complexity, 24, 33, 59

junk DNA, 52, 54,

kinase, 68-69

kinetechore, 59, 63

lamella, 41

lamellipodia, 77, 80

lysosome, 42-43

meiosis, 59-64, 80, 104

messenger RNA (mRNA), 14, 29, 32, 39, 42-43, 45, 50-52, 54, 79

microtubule, 45, 59, 63, 75-77, 80

Miller-Urey, 15-17

mitochondria, 25, 27, 40-42, 45, 76, 80

mitosis, 56-60, 62-64, 76, 80

module, 40

Mus musculus, 37

Mycoplasma genitalium, 18

nucleoid, 28

nucleolus, 39

nucleotide, 13, 18, 49, 63-64

nucleus, 25, 27, 33, 38-39, 41, 43, 45, 50, 52, 54, 58-60, 63, 68, 92, 104-105, 110

operon, 54

peptidoglycan, 28

peroxisome, 43

phagocytosis, 42

photosynthesis, 41, 83-85, 87

pili, 29

plasma membrane, 13, 24-25, 28, 41, 43, 60, 67-68, 77-79

plasmid, 29

prokaryote, 25-30, 33, 37-38, 43, 45, 50-51

proteinoid, 17

racemic, 16-17

recombination, 61, 63-64, 113

replication, 13, 37, 49, 57-58, 64

respiration, 41, 84-85

ribosome, 14, 29, 32, 39, 43, 50, 52, 54

RNA transcript, 14, 37, 41, 51

Saccharomyces cerevisiae, 37

somatic cell division, 58

synapse, 65, 70-71, 78

transcription, 37, 39, 49, 50-52, 54

translation, 37, 43-44, 50, 52-53

trophoblast, 101, 107, 110

tubulin, 75-77, 80

vacuole, 42-44

vesicle, 53

zygote, 95-97, 101, 103-104, 107, 110